RICHARD DANNE

SHOOTING FOR THE STARS

Six Decades of Timeless Design

For Marty,
a great colleague and friend

Dick

ORO
EDITIONS

ORO Editions
Publishers of Architecture, Art, and Design
Gordon Goff: Publisher

www.oroeditions.com
info@oroeditions.com

Published by ORO Editions
Copyright © Richard Danne and ORO Editions 2024
Text and design Images © Richard Danne 2024

Graphic Design: Richard Danne
Full Credits on pages 246-247
ORO Managing Editor: Kirby Anderson

10 9 8 7 6 5 4 3 2 1 First Edition

Library of Congress data available upon request. World Rights: Available

ISBN: 978-1-961856-13-4

Color Separations and Printing: ORO Editions, Inc.
Printed in China.

International Distribution: www.oroeditions.com/distribution

ORO Editions makes a continuous effort to minimize the overall carbon footprint of its publications. As part of this goal, ORO Editions, in association with Global ReLeaf, arranges to plant trees to replace those used in the manufacturing of the paper produced for its books. Global ReLeaf is an international campaign run by American Forests, one of the world's oldest nonprofit conservation organizations. Global ReLeaf is American Forests' education and action program that helps individuals, organizations, agencies, and corporations improve the local and global environment by planting and caring for trees.

Dedication

For those people who have provided inspiration and motivation for this book:

My beloved parents Helen and John;
Dear Barbara, personal and business partner;
Gifted writer aunt Mary Agnes Thompson.

"You should feel gratified that your work resonates
with so many people after all these years.
But, of course, that's what it means to be timeless."
Michael Bierut, designer/partner, Pentagram/New York

Contents

Foreword

This book is a style guide to designing for the future. My career is still in motion after more than six decades of successful design programs, involvements, and victories.

I've been blessed to have known and worked with many corporate leaders and with the giants of design – the ones who invented visual communications as we know it.
We all can agree, lasting design is good communication as it adds substance and value to an organizations' efforts.

I've also had the privilege of leading our professional organizations to elevate the space for all to practice.

Now it's time to focus on the next generation. And here, I am doubly blessed to have connected and shared with many younger professionals.

My lengthy career has been dedicated to *timeless design*, that which endures well into the future. I am optimistic this book can be useful to the next great wave of leaders, communicators, and designers.

Colleague Quotes

He's a leader, gentleman, and a one-of-a-kind shining star in our vast design galaxy.
Dana Arnett, founder and chairman, VSA Partners, Chicago, IL

Richard shows the power of design to make the world a better place, and he works tirelessly to spread that message to the next generation of aspiring designers. He is a constant source of support and inspiration.
Kaelig Deloumeau-Prigent, founder & co-chair, Design Tokens Community Group, Seattle, WA

Your creative memoir is simply stunning... as in being stunned with the elegance, thoughtfulness, appropriateness, and relevance of your life's work. It triggers all of those stirrings that made my career in design so invigorating and meaningful.
Ric Grefé, former AIGA executive director, Williams College, Williamstown, MA

There are very few designers that have helped create the visual culture of our time. Richard Danne has not only designed some of the most recognizable logos of the 20th century, he has done so with a sense of whimsy, aplomb, gravitas and ingenuity utterly and uniquely his own. All designers everywhere owe him so much.
Debbie Millman, writer, educator, artist, designer, host of the podcast "Design Matters," New York, NY

The 30-year relationship between FIT and your firm has been nothing short of extraordinary. You approached every strategic initiative as well as every short-term project with a fresh approach and creative intelligence.
Loretta Lawrence Keane, former vice president, communications and external relations, Fashion Institute of Technology, New York, NY

Congratulations for a thoughtful, intelligent, insightful commentary on design in our culture. I know the interview was about your work, but it spoke volumes about the best in our field today. Thanks for carrying the banner for us all.
Kit Hinrichs, founder / creative director, Studio Hinrichs, San Francisco, CA

Richard Danne is a design visionary who has worked on some of the most influential design projects of the past half century – while he also served as the U.S. President of Alliance Graphique Internationale, an elite organization.
Rob Auchincloss, publisher, HOLOCENE magazine + podcast, New York, NY

For over 60 years, in a world of shifting graphic styles, evolving attitudes, and radical changes in technology, Richard Danne has been a steady beacon of talent, intelligence, warmth, and integrity.
Tom Geismar, designer/partner, Chermayeff & Geismar & Haviv, New York, NY

It all goes to prove what we felt from the beginning: we picked the right firm.
J. Murillo Valle Mendes, President, Mendes-Junior, Belo Horizonte, Brazil

Richard Danne's luminous intelligence, voracious curiosity, and lovely human warmth animate a body of work that – for more than six decades now – has been one of the most powerfully influential in American visual culture. He exemplifies the very best of what is possible both in design and in the art of life.
Lana Rigsby, designer/partner, Rigsby Hull, Houston, TX

For years, it's been my good fortune to have kept up a wonderful personal relationship while delivering excellent professional work.
Chris Conover, then Mngr. international communications, Pratt & Whitney, Hartford, CT

You have to know that in my early years, you were one of my design idols, and you still are.
B. Martin Pedersen, publisher/designer, Graphis, New York, NY

We all had the privilege of hearing Richard Danne share his prolific design career as well as his incredible experience with AIGA. It was both inspiring and delightful.
Jina Anne Bolton, presenter of both: Clarity Conference and Salesforce + AIGA/SF events, San Francisco, CA

Your book is great, it's like a design time capsule.
Dr. Gjoko Muratovski, Director, Digital Futures, Deakin University, Melbourne, Australia

It would appear that you double dipped in the creative line, getting a full scoop of pro designer extraordinaire, then going back for a full helping of musical talent with a Steinway topping.
Craig Frazier, designer/illustrator, Mill Valley, CA

You did a tremendous amount of work organizing the AGI Montauk Congress. We had a wonderful time and, from my point of view, everything went like clockwork. Thank you! Thank you! Thank you!
Fritz Gottschalk, designer/partner Gottschalk + Ash, Zurich, Switzerland

Dick Danne is the quintessential designer, AIGA Medalist, and mentor. His work at NASA is an epic story of visual invention and what it means to bring an idea to life.
Julie Anixter, then AIGA Executive Director, New York, NY

The students continue to talk about your visit and work. Your input and lecture changed how they see branding and design.
Sean Adams, Director, ArtCenter College – Graphic Design Graduate Program, Los Angeles, CA

Never meet your heroes, the saying goes.

Ignoring that advice, in 2014 we timidly composed an email to one of our design heroes, Richard Danne. We'd never met and, as much as we hoped, we didn't really expect a reply.

Earlier that year, we had published a reissue of Unimark's 1970 NYCTA Graphics Standards Manual using Kickstarter and, in doing so, established our indie publishing imprint, Standards Manual.

To us, Danne & Blackburn's 1975 NASA Graphics Standards Manual was the holy grail of design manuals. But like the NYCTA manual, original copies of the NASA manual were rare and the full guidelines were not available online. So, we decided to go for it and see if Richard would be interested in chatting about publishing a book. A real moon shot!

Dick replied quickly: "I was wondering when I'd hear from you guys."

That was the start of an incredibly fulfilling partnership, a lot of work, and most importantly, a lasting friendship. After publishing the manual reprint in 2015, it has been thrilling to see "the worm" and the entire NASA design system roar back into global culture and – most excitingly – on NASA rocket hardware, where it belongs!

We are humbled to be a small part of Dick and Barbara's lives. As you can see on the following pages, this performance goes far beyond the NASA project for a beautiful, lasting, meaningful body of work!

Thank you Dick for being our mentor, friend, and hero.

Hamish Smyth & Jesse Reed, partners, Standards Manual, Brooklyn, NY

*William Widmaier (hat) with some of
his global team, after Richard's presentation,
Norton Design Summit, Palm Springs, CA*

One of the great design sagas of this or any age. It embodies achievement, loss, rebirth and renewal – a singular experience over my expansive career.

The NASA Odyssey

"The Boys"

Bruce Blackburn (left) and Richard Danne,
photographed in the new Danne & Blackburn
Studios at One Dag Hammerskjold Plaza,
New York City in 1975.
This image was taken soon after the release of

Our NASA redesign program had a spectacular ride,
then suddenly it was gone.
Yet decades later, it came roaring back to life!

It was October 1974, and my partner Bruce Blackburn and I were headed to Washington, D.C., for our initial design presentation to NASA administrators. The National Endowment for the Arts' "Federal Graphics Improvement Program" was quite new, and the NASA redesign was both an early and important test.

There were very few NEA guidelines for this phase of work, but we had really knocked ourselves out... going the extra mile. Danne & Blackburn had invested an extraordinary number of hours on research, design development, and the presentation. Why would we do this? Not just because we wanted to look good – we did – but because we felt pressure to gain a quick and vital victory for the NEA. No doubt about it, a lot was riding on this one.

After the successes of the Mercury and Apollo programs, NASA now found itself in a slump, impatiently waiting for their Space Shuttle program to kick in. There were no automatic headlines for the agency now. Also, though the NEA had convinced NASA to embark on the redesign, the agency wasn't obligated to go the distance. Most of the federal agencies that signed up for redesign were just as cautious. Agreeing to a Phase 1 study didn't mean a commitment to implement the study's conclusions – it just meant, "Let's see what you've got."

Many of us who worked on these massive federal redesign projects were surprised that all of this was happening on Richard Nixon's watch, but the NEA's Nancy Hanks was the real force behind it all.

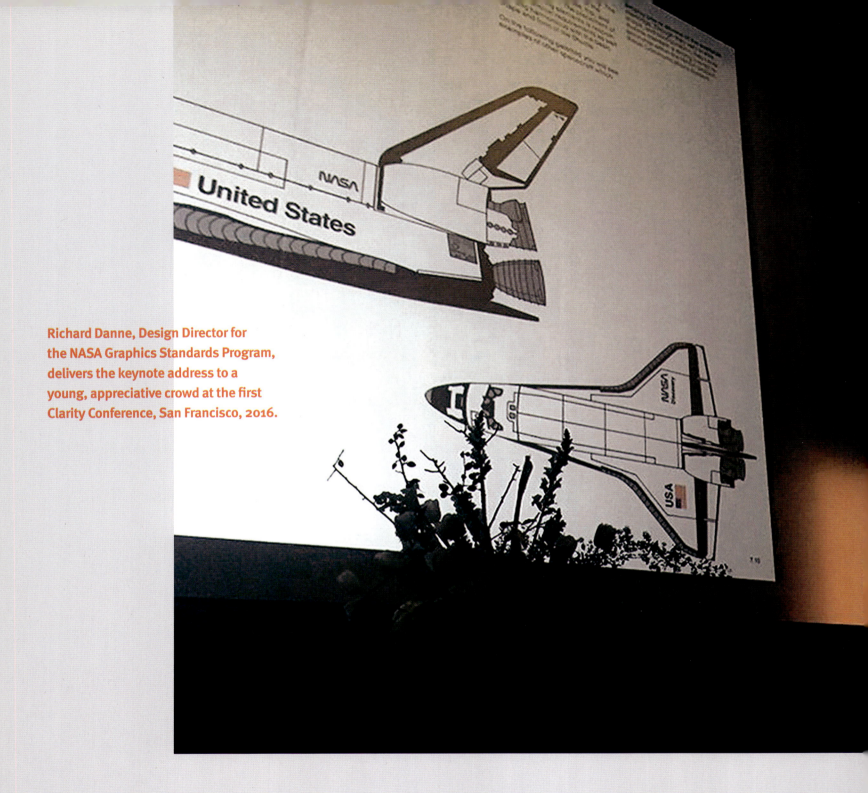

Richard Danne, Design Director for
the NASA Graphics Standards Program,
delivers the keynote address to a
young, appreciative crowd at the first
Clarity Conference, San Francisco, 2016.

*Shown are some of
the 25 demonstrations
from our 1974
design presentation.*

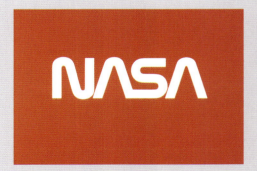

Dr. James Fletcher was the NASA administrator at the time, and his Deputy was Dr. George Low. Several other staff members would attend the closed presentation, but it was clear the show was for these two individuals. George Low had actually served as interim administrator, prior to Fletcher being selected for the administrator's spot, adding some intrigue to the relationship.

After the proper introductions, our 35mm slide presentation got underway. One objective was to make a case for replacing the NASA insignia (nicknamed the "Meatball") with a more useful new logotype. The "Meatball" was complicated, hard to reproduce, and laden with "Buck Rogers" imagery. Clearly it was born out of the classic airman syndrome where hype and fantasy dominated over logic and reality. Our logotype was quite the opposite: it was clean, progressive, could be read from a mile away, and was easy to use in all mediums – it later survived much of the inferior printing from the GPO (U.S. Government Printing Office) of that time.

It is important to note that Danne & Blackburn (D&B) was only presenting one solution to the client; it was a controversial idea for the time as it would be today. But we had developed a nomenclature system with the various center names all linked to headquarters in a totally democratic way. Yet the logotype was radical enough that that the room was abuzz. We trotted through the various applications we had designed to show how strong and effective the program would be when it was fleshed out. Though it was not required in our contract, we had decided that it was the only way to make the point: This is a coordinated, comprehensive design program, not just another ornamental badge to be stuck on a multitude of products by countless personnel and sub-contractors.

These attendant visuals went a long way towards making the entire program "real" and convincing. As the presentation unfolded, there was considerably more acceptance of the program. Perhaps all the hours invested were worth it.

Before

After

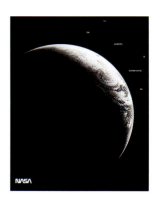

But there were residual issues, and the focus shifted back to the logotype itself. And here is one of the most interesting exchanges I've ever witnessed in a design presentation:

Fletcher: "I'm simply not comfortable with those letters, something is missing."
Low: "Well, yes, the cross stroke is gone from the letter A."
Fletcher: "Yes, and that bothers me."
Low: "Why?"
Fletcher: (long pause) "I just don't feel we are getting our money's worth!"

Others, not just the designers were stunned by this last comment. Then, the discussion moved back to the strong red/rust color we were proposing. We had tried many other colors, of course, including the more predictable blue range but settled on red because it suggested action and animation; it seemed in the spirit of this *"can do"* national space agency.

Fletcher: "And this color – red – it doesn't make much sense to me."
Low: "What would be better?"
Fletcher: "Blue makes more sense, Space is blue."
Low: "No Dr. Fletcher, Space is black!"

This is not to suggest there was any animosity between the two men, and I doubt if there ever was. But, their dialogue is still pleasantly stored in my memory bank.

We didn't get the ultimate win that day, as no one signed off on the presentation. But we left feeling our chances were far better than 50/50. Bruce and I headed back to New York on the Eastern Shuttle and had to simply wait for the NASA response. The call came within a week from Jim Dean, our contract coordinator at the agency – and our small firm was thrilled to get the news... "It's a go!

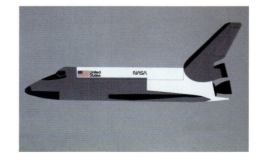

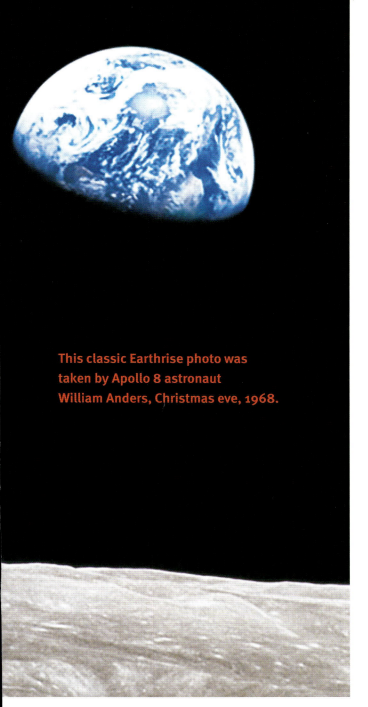

This classic Earthrise photo was taken by Apollo 8 astronaut William Anders, Christmas eve, 1968.

At times I wish the story had ended right there with a win and those feelings of optimism. But something got lost in the transition.

The agency decided to alert the various centers of the new program by sending Executive stationery to each center director. That stationery displayed the new NASA logotype and it was the first time they were informed of the graphics program and image change.

NASA was a coalition of many different agencies that had been operating independently for decades. Formerly known as NACA, the agency was rebuilt and renamed the National Aeronautics and Space Administration on July 29, 1958, by President Dwight Eisenhower. These centers were like fiefdoms, they enjoyed their freedom and their provincial specialties. At the time of our redesign, there was resistance to almost anything emitting from headquarters. The centers were like unruly children... and to say they were competitive would be an understatement.

So, this is the environment we confronted. Those letterhead "gifts" coming out of Washington and sporting the new logotype proceeded to detonate across the country – and all hell broke loose! Our firm wasn't keen on introducing the whole program in such a shallow and casual way. For one thing, the letterhead couldn't explain how thorough and solid the new graphic system was.

Too late, headquarters realized they had made a mistake and needed a solution. What ensued was one of the most difficult assignments ever. Along with a Public Affairs representative from Headquarters, we would travel around the country from center to center, and give that full design presentation... over and over to somewhat hostile audiences. Often they were "loaded for bear" before we arrived. But – as awkward and tiring as it was – it had to be done.

Chris Kraft, a NASA notable and the "Voice of Mission Control" for

Photo left: Launch photographers at Cape Kennedy in the early 1980s.

many years, was now the administrator of Johnson Space Center. His career went back to NACA, and he was hot under the collar when I arrived in Houston. But after viewing the lengthy presentation, he said: "Why wasn't it handled this way from the start? I can see it's a real program, and I'm OK with it now."

Whew! It took months to smooth ruffled feathers and build a base of trust with the Centers. Of course, the old guard was bitter and didn't want to give up their beloved "Meatball." They continued to employ subterfuge, to undermine the redesign effort wherever possible. They coined the term "Worm" for our Logotype. It was meant to be derogatory but it also became one of endearment over time (not unlike "Meatball"). Younger NASA employees strongly preferred the new graphics program and a schism developed in the agency: old vs. young.

Over the next months, D&B produced the most basic graphics standards materials, about 30 pages worth, including all repro art and custom color chips for each center. Also, on our recommendation, NASA hired an internal graphics coordinator to monitor the whole program. Bob Schulman was ultimately chosen and, though he was never shown the logotype or design manual during his interview, he became a staunch supporter and valuable ally for many years.

The manual continued to evolve over the next months, and it now included a strong endorsement intro by Dr. Fletcher. In the end, it reached about 90 pages and covered every aspect of NASA: ground vehicles, aircraft, signing, uniform patches, publications of every kind with font guidance, business forms, public service TV titles, the Space Shuttle, other space vehicles and satellite markings.

The Presidential Design Awards were established to recognize and honor the best of those NEA Federal redesign efforts. 1985 was the first year of the awards, and our NASA program was singled out with the "Award for Design Excellence." In ceremonies held in Washington, DC, on behalf of Danne & Blackburn, I accepted the award from Ronald Reagan.

The firm of D&B was dissolved in January of 1986. Then I launched Richard Danne & Associates, retained most of the staff, and the accounts which I had handled over the years – including NASA (it was largely inactive by then). But in 1992, I was told that the new administrator Dan Goldin was landing at Langley Research Center which had a large red logo on the building. A couple of older staffers touring with him made some disparaging remarks about "the worm" and, according to witnesses, Golden asked: "Can I change that?"

A couple of weeks later I took a call from Bob Schulman, and he relayed that Dan Goldin story to me. He then dropped an incredible bombshell – our NASA design program had been rescinded, and the "Meatball" was being reinstated. He then asked if I would want to handle the retro conversion. It took a nanosecond to respond: "Absolutely Not!" Schulman, the effective "gatekeeper" for our program was also crestfallen, and he uttered: "I knew that's what you would say, but felt I had to ask."

As professionals, we are used to setbacks and disappointments. Often things happen with a change in management, a new marketing team, or an election delivers a new administration. It's all part of the game and there is little you can do about it. But reflecting back on the NASA program, it was quite painful. My solution was to not think about it, just keep moving forward.

left: The first of our three U.S. Presidential awards, for the NASA design program, January 30, 1985.

Easier said than done! Our design program is undoubtedly one of the most published in the world. Decades after the program was scuttled, I continually filled requests for editorial background and visuals from that 1975 Graphics Standards program.

One European publication profiled NASA in this way: *Graphic Classics:* "Icons of graphic design that have set a benchmark for excellence and innovation."

While London's *Creative Review,* in a worldwide competition, judged it among the " Top 20 Logos of All Time."

Countless international graduate design students requested help as they wrote and designed projects focused on our program.

The famous British Design critic Alice Rawsthorn penned an astute piece for the *New York Times Magazine.* Her bottom line premise? "When it comes to logos, the space race is still alive."

And in a major 2015 development, D&B's original NASA *Graphics Standards Manual* was reissued as a record-breaking, wildly successful hard cover book.

Then to cap it off, in 2022, our popular logotype was finally and formally reinstated into NASA's design vocabulary. Hallelujah!

As Michael Bierut, noted designer and author, stated when closing our super interview at The Strand Bookstore in New York City: "A design program this elegant, this good, really can't be rescinded or taken away. Its power is as evident today as it was when first presented to NASA on that slide projector back in 1974."

Well spoken, Michael, and thank you publishers worldwide.

Graphics Standards Manual, opposite, with pages created between 1975 and 1980.

The "sold out" NASA Interview at Strand bookstore in NYC, with Michael Bierut.

Coming Home!
A fulfilling lecture at NASA's Jet Propulsion Laboratory, Pasadena.

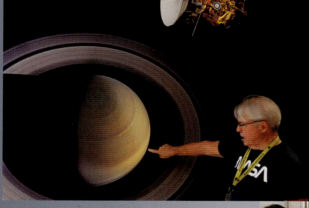

David Doody, Cassini Realtime Flight Operations Lead Engineer shows his long-time support with this handsome T-shirt.

Salesforce / AIGA, San Francisco.

Lectures at the ArtCenter Graphic Design Graduate Program, and Hoffmitz Milken Center for Typography, ArtCenter, Los Angeles.

Rebirth + Redemption

In late 2015, the original *NASA Graphics Standards Manual* was reissued as an exceptional hard cover book by the stellar team of Jesse Reed and Hamish Smyth (of Pentagram NY, now at Order). Its 220 pages include: The entire manual, essays, the original presentation, plus other supplemental material. A dramatic success, the project was crowd-funded by 8,798 backers to the astounding tune of $941,966. Our dormant design program came roaring back to life! Now in its seventh printing, the impressive book is available online at: www.standardsmanual.com

Yet another *NASA Manual* reissue has been published by two talented French designers, Sacha Léopold and François Havegeer. Their handsome, more modest version is translated in French and used for teaching in the schools of France.

And I've been lecturing and podcasting extensively, retracing the exciting NASA design program experience. Audiences tend to be rather young and often describe the story as "inspirational."

Everywhere you look, there are logos and fonts influenced by our NASA program – it has more legs today than when it was first introduced in 1975.

April 2, 2020: On my birthday, SpaceX and NASA announce they will revive the logotype for their DM2 Dragon Rocket as it returns two Astronauts to space.

Logo apparel is especially popular with young customers.

+Renewal

The Orion spacecraft will feature the logotype on future Artemis missions.

SpaceX officially returned American astronauts to space on May 30, 2020. I knew something was up when I took a call asking me to confirm the "NASA Red" color. It wasn't much, but it was a tipoff they had some adventurous ideas up their sleeves. Since NASA has involved commercial developers and contractors in their endeavors, SpaceX has been perhaps the most innovative partner of them all. Their decision to apply our logo to the *Crew Dragon* rocket opened the door to possibilities with other contractors as well as with NASA itself.

Sculpture at Headquarters entrance, Washington, D.C. Installed June, 2023.

The most interesting part for me personally? After some 35 years, I've had a consulting agreement with NASA once more. And under the leadership of Bert Ulrich at Headquarters, the 2020 *NASA Graphics Standards Manual* features our logo again. I'm thrilled to be around for the comeback!

Fashion note: NASA sells a wide variety of apparel featuring our very popular "worm" logo. Coach, Vans, and H&M have been involved, to name a few. These and others have enjoyed robust international sales, and applications to use the logo have surged.

So the D&B program never really went away – at five decades later, it is vibrant and alive. Rebirth and Renewal!

Copy Cat Culture!

A popular segment of my lectures and podcasts features the pervasiveness of logos and fonts clearly influenced by our NASA logotype. And this is especially true if the subject is space, flying, technology, or most anything about the future!

Here is a random selection from this look-alike phenomenon which audiences enjoy.

General Motors was the very first to use cloning for a marketing edge.

This great musician used all the letters for his video title.

The 2018 Olympic snowboarding team with space-theme apparel.

Virgin America refreshment cart on a flight to New York.

The new high-tech electric bike from Brooklyn-based Tarform.

Play Station looks ahead with style.

Still prominent in the American landscape, here a Kiddie Kar ride at a mall.

Driving in our hometown when a van pulls alongside with this window sticker!

Bold, innovative electric delivery van arriving from the UK.

"The Martian" blockbuster movie with its ARES III logo font.

NASA's own online retail shop for space memorabilia.

The newest uniforms for the "Space City" Houston Astros baseball team.

The Boeing Leap logo on an enormous engine for its "Dreamliner" aircraft.

San Francisco's Betabrand apparel sports a familiar look on their "space jacket."

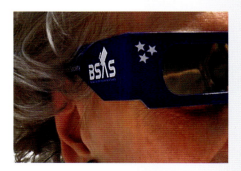

During a recent Solar Eclipse, protective glasses with similar font.

November 6, 2023: Events at NASA Headquarters celebrating the comeback of an iconic logotype from 1974.

The rewarding day began with a panel discussion, streamed from D.C., about "the Worm" – its history, longevity, surging popularity. Left to right: Moderator Shelly Tan (Washington Post), Danne, Michael Bierut (Pentagram / NY), Bert Ulrich (NASA Branding), Julie Heiser (Amazon Music).

Former NASA astronaut and associate administrator Bob Cabana awards the Exceptional Public Achievement medal.

"NASA employees and visitors assembled outside, to dedicate a sixteen-foot-long rendering of the worm in front of NASA HQ and to present Danne, who is eighty-nine, with its Exceptional Public Achievement Medal. A number of young NASA designers from the Goddard Space Flight Center, in Maryland, crowded around the worm and chatted with Danne. When it was all over, he took a breather in the NASA library, amazed at his own trajectory, from a Dust Bowl Oklahoma farm to this splendid comeback, his design again heading for space. "When I first came here, in 1974," he said, "I thought this was heaven."

Excerpted from Bob Sullivan's article in **The New Yorker,** Nov. 27, 2023

*Call it Leadership, or call it Service,
it's part of my DNA. My parents
challenged me to "Be the best you can
possibly be, and share with others."*

AIGA Immersion

"As a designer, Richard Danne can lay claim to an outstanding body of work. And, as a citizen of New York's design community, Dick contributed generously of his time and talent. He served as president of AIGA at a time of critical transition and guided the organization from a largely local club to a national institution."

George Tscherny, former AIGA President and Medalist.

When starting my independent career back in Dallas,
I hoped to be involved in U.S. design leadership,
*and help make **"change."** Be careful what you wish for!*

The American Institute of Graphic Art (AIGA) has a rich tradition. It was founded in 1914 at the National Arts Club, 119 East 19th Street in New York City, less than a block from my old Gramercy Park office. Typography, book design, book binding, and printing was the early focus of the Institute, and the prestigious AIGA Medal was created and awarded for the first time in 1920. The term "graphic design" would not be coined for many decades, and it wouldn't be until the design and communications of the 1950s and 60s – not the trades – would drive the organization.

I was already a national member of AIGA upon arrival in New York in 1963. Getting involved in any organization is pretty much the same: you start at the bottom, volunteer for the mundane donkey work, and gradually work your way up.

I helped install several exhibits – mostly manual labor. After paying those early dues, Phil Gips and I designed and installed a packaging show. Some years later, I chaired the 1974 "Communication Graphics" show, one of their biggest. I assembled a national jury, and after the opening, I got a letter from Executive Director Ed Gottschall saying, "This year's CG process, through last night's opening, represents the very best that AIGA has to offer. Thanks for your superb efforts."

Ed retired in 1976 and was replaced by Flora Finn Gross. Over the next year, and for many reasons, the AIGA would experience a downhill slide and sink into substantial debt. I had just been elected to the board and witnessed a leadership vacuum at the top and saw there was little exchange or communication (definitely not all Flora's fault). And though some exhibit openings were elegant, they didn't make sense when we were slipping further in debt each day. Clearly something had to be done to avoid a crisis.

I was judging the Mead Annual Report Show at a midtown hotel when a phone call came from the chair of the Nominating committee for the upcoming slate of AIGA officers. He went right at it, "We are asking you to be the next President." I recall backpedaling, suggesting others like Tom Geismar, whom I admired, but Tom was on the committee so he couldn't be nominated. Obviously, this was far too big a deal to answer on the spot. I promised to return the call within 24 hours but was somewhat dazed for the rest of the judging.

I desired a role in American design leadership, but these were far from ideal circumstances. I was young, it would be a colossal job and a huge can of worms. I talked it over with my personal partner Barbara and my business partner Bruce. Someone has to do it so, with mixed emotions, I phoned back the following day, "OK, let's do it."

I had some quick sessions with Flora and reviewed the status more closely. It was all very convoluted and I soon realized we were in more trouble than anyone realized. While I liked Flora personally and we got along well, it was obvious that she was quite distraught and in over her head. Everyone on the executive team knew something had to be done, and we decided on a change sooner rather than later. We would simply have to limp along without an executive director until a replacement was found.

AIGA 2014 Centennial Medal.

The word went out, and the committee started initial interviewing. I received a call from Caroline Warner Hightower – who sounded interesting on the phone – and we agreed to meet the next day. She came to my office around 11 am, and we closed the door and proceeded to talk. It was clear she was a good fit, plus she actually knew what design was... and liked it!

We simply "hit it off" in a way that rarely happens. Caroline asked if there was anything she could do to help her get the job. I suggested that a couple of letters of recommendations from her professional past would help. She said these letters would be available as soon as possible, and they were. The letters only confirmed what I already knew in my gut: I needed a talented teammate, someone with the right instincts for the design community, and a kindred spirit. I've always loved serendipity. Caroline happened along – the right person.

The committee met with Caroline and agreed, of all the candidates, she was the best qualified to help lead us forward. Done!

My job just got better. It would never be easy, but at least it could be doable. And one other thing, it's not always a disadvantage to take over an enterprise when it's in trouble and in decline. Change is possible when things are not right, when an organization is floundering.

When I was younger and practicing in the hinterlands, many of us felt a little left out. Everything was created in New York and emanated from there. AIGA was headquartered in NYC, the board members were all New Yorkers, and the exhibitions opened there – some traveled, many didn't. Lectures and social events were there, and it was great... if you lived in New York. Essentially AIGA was a wonderful New York Club, and had not fully recognized the need to become a national entity.

And here is the landscape we had just inherited:

National membership

1,600 members and decreasing.

Put in perspective, AIGA was the size of today's San Francisco Chapter!

Chapters

Two, yes, two.

Cleveland, where Bert Benkendorf was a one-man whirling dervish and bitter at headquarters for lack of communication.

Washington, D.C., an almost dormant chapter with few members, just as disenchanted with headquarters and in a sorry state of disrepair.

Financial status

We were in debt – no pretty way to portray it. We owed money to printers, binders, caterers, shippers, utility and insurance companies, and the landlord. Even if the executive director wasn't up to the task, clearly the officers and treasurer had been nodding at the wheel. Years later at Caroline's retirement dinner, I commented on our plight, "We first tried paying our bills with orange Monopoly money, when that didn't work, we took to printing our own money!"

Publishing

Almost nothing. A sometime periodical would go out but not much else.

It was dire enough that I called for a two-day Crisis Retreat in the woods: the entire board and notable people from Chicago, D.C., and elsewhere (no phones or distractions). It was serious, well attended, and yielded the desired results. And to stay engaged and solve the problems, the board met every single month!

So long story short, we did rally the leadership and right the ship. It wasn't easy, but here are my comments on leaving office in 1979:

My remarks on leaving office, AIGA Annual Meeting, July 27, 1979

"When I accepted the office of President on July 28, 1977, I stated certain goals that I felt we should strive for at AIGA. I would like to quickly give you a status report on those goals.

1 Fiscal Responsibility

- We were substantially in debt. We are now solvent.

2 Education and Resource

- We have just installed the beginnings of our library and archives.
- We've assembled a listing of all graphic design courses offered at the various schools and institutions around the U.S.; this will be published shortly and will be available to all young people interested in pursuing a design career.
- Under Chairman George Sedak of Cooper Union, the Education Committee is planning a series of activities, including a student exhibition in spring of 1980.

3 Membership

- We are now launching a nationwide membership drive, a campaign many months in planning. This effort is chaired by Dick Lopez .
- We are publishing immediately an AIGA Membership brochure and AIGA Directory. The Directory will include names, addresses, and a description of each member's design activity – presented by geographic regions.
- We are regularly publishing an AIGA Newsletter.

4 National Participation

- We now have board members from Los Angeles, San Francisco, Chicago, Boston, and Washington, D.C.
- We are opening in Chicago next spring, the AIGA Packaging Show. This is our first exhibition to be chaired, managed, designed, and opened in a city other than New York.
- Andy Kner heads a study group which is investigating the advantages and disadvantages of further AIGA chapters, an issue which has long needed res - olution and clarification.
- AIGA projects have been designed, typeset, and printed in many U.S. cities.

5 Publishing

This is the heart of an effective, outwardly-focused AIGA.

- In the summer of 1980, AIGA will introduce, through a major U.S. publishing house, a 400-page case bound volume entitled *Graphic Design USA #1*. It will record in 4-color, the AIGA exhibition season, including the Medalist, and act as the definitive book on design in America. Excellent work has been done by the committee co-chaired by Bob Bach and Martin Moskoff. (Committee: Martin Fox, Bob Scudellari, and Tom Suzuki).
- We have researched and written for the NEA, a paper entitled, "The Cost Savings Benefits of Design Systems" by Caroline Hightower.
- We will publish shortly, through another publishing house, an essay entitled, "Graphics Guide for Non-Profit Institutions." This project designed and produced by Massimo Vignelli, will benefit countless institutions across the U.S.

6 Program

- NEA meetings of the following American Design Societies and Institutes: AIA, IDSA, AID, and AIGA to strengthen public recognition of designers and design excellence.
- We are anticipating a grant to produce a 30-minute film of "Film Graphics" for airing on public broadcasting as well as multi-purpose use.
- We are cooperating with the Library of Congress to develop a bibliography of graphic design.
- We are now the medium through which the NEA selects and reimburses its panelists and consultants.
- The latest phase of the Department of Transportation's Symbols/Signs effort, under the direction of Tom Geismar, has been just completed and will be published as a major document in the next months.
- Our new Enlightened Client Award is being chaired by Bill Bonnell. Hopefully, it will be an event for spring of next year.

Needless to say, it has been a very busy two years!

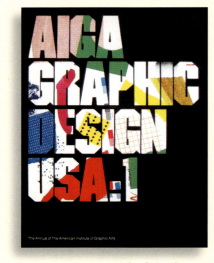

The first edition, jacket design by Tom Geismar.

We at AIGA are blessed with a wonderful staff. I would like to thank Susan Marks, now residing in Paris – who helped me greatly before we had the good fortune of hiring Caroline Hightower as executive director.

To Caroline, Deborah, Nathan, Victoria, and Shelley, I say – well done!

I am thrilled to be leaving office – to be hanging up my spikes – with so much optimism for AIGA. Over these next years, with the new additions to the board, with Caroline and her spirited staff, I can afford to be confident for AIGA.

Thank you."

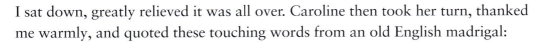

I sat down, greatly relieved it was all over. Caroline then took her turn, thanked me warmly, and quoted these touching words from an old English madrigal:

"Where beauty moves
  and wit delights
  and signs of kindness bind me,
  There, where'er I go
  I leave my heart behind me."

"These words certainly expresses the feelings of all of us here.
  You brought decency and expertise to this situation..."

The New York Chapter still loomed ahead, but it was not an immediate concern.

Birth Of A Chapter (excerpted)

I'm truly sorry I can't join you tonight, I'm deeply involved in a Jazz Festival in Napa Valley this week. But, wow! The feelings come flooding back. While it seems like only yesterday, it actually happened... in another century!

Steff Geissbuhler brilliantly delivers my speech at the opening event staged at SVA.

A few facts about that earlier landscape, before the dynamic New York Chapter was born:
- I served as president of AIGA National from 1977 to 1979;
- Other design organizations were attempting to grab the national spotlight and plotting to become the true National Institute for Design;
- I and others felt the AIGA was still our best hope, but it needed to grow and represent ALL regions of the country. To accomplish this would require outreach and many more chapters. At that time, I envisioned a reasonable goal of 10,000 members;
- And if this concept was to gain traction, there would have to be a New York Chapter: for symbolic reasons; to demonstrate change; and to help lead the expansion charge.

I left the presidency on a positive note, fully expecting progress on chapters and that plan for national growth. But something got lost in translation. After two years, and frustrated with a lack of progress, I decided to generate AIGA/NY myself. I enlisted a good mix of young professionals for a large organizing committee, the next generation of talented, forward-thinking designers – Ken Carbone, Louise Fili, Michael Donovan, & many others. We met often and worked hard, writing chapter by-laws from scratch, creating program ideas customized for the NY design community, recruiting financial support, searching for venues to stage events, organizing fundraisers, acquiring non-profit status in NY State, dealing with bank accounts and all the necessary nonsense.

It took most of a year, headquarters wasn't overly helpful, and our committee got very restless. But, it was essential to have a NY Chapter to show that the AIGA was serious about inclusion, outreach, and national leadership.

Yet, AIGA/NY became a reality in 1982. We came out of the gate with a full blown 1983 program which included:

"Richard can be considered the father of
AIGA Chapters."
Sean Adams, former AIGA President and Medalist

- Features on "The American Magazine;"
- An ongoing series of ethics and professional practices by Roger Whitehouse;
- The "idea exchange" series with notable speakers from NY and beyond by Peter Bradford;
- A continuing series on the history of graphic design by Keith Goddard;
- An innovative film and motion graphics series by Richard Greenberg; and more.

We begged for various venues for these events. F.I.T., one of my long-time clients, helped immensely and we used their facilities over and over. It's a blur, but my 1983 - 84 Report to the National Board stated that AIGA/NY had staged an event almost every two weeks. Whew!

But, we were becoming a model AIGA chapter, animated and productive, while sending a very strong signal to the rest of the country. And, headquarters was better prepared to serve a larger national design community in a more balanced, democratic way.

My original commitment to AIGA/NY was for one year. But two years had elapsed, and I was experiencing some burnout. And my own practice was calling, *loudly!*

But wait, there was light at the end of the tunnel. All we needed now was a super person to lead the chapter forward for the next couple of years – Steff Geissbuhler was the perfect choice. It still was "the job from hell" but I offered Steff several persuasive reasons for accepting it:
1. It would look swell on his resume and enhance his profile in Gotham.
2. It was really important work.
3. He could lick all those postage stamps going on our chapter mailings!

Done. Steff bought in, the new Board was elected, and everything else is chapter history. That group put together two great years of programming which, you'll hear about next. My guess is that Steff can still taste that glue, but he must have forgiven me because we're still close.

And looking ahead, I'm quite sure of one thing: the next 30 years will be very different! I'm equally sure that AIGA/NY is going to be your best friend through these uncharted waters.

Warmest congratulations to the New York Chapter, the dream came true so I'm feeling proud. AIGA/NY will always be the most important chapter not only for its size or talent but for its seminal role in the evolution of AIGA.

And now to pass the baton to Steff, again. Thanks pal – and onward!

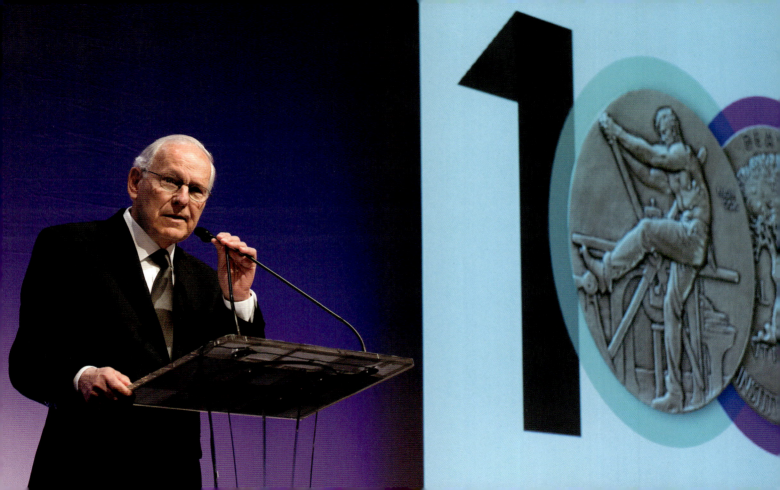

Richard Danne, AIGA Centennial Medalist

By Margarethe Laurenzi

Richard Danne has lived and practiced design by a singular principle: Create work that has beauty, simplicity and permanence. A classically trained graphic designer, Danne has applied a fresh, contemporary sensibility to every project he has undertaken – from his signature corporate annual reports, to a total program for NASA that remains one of the world's most-recognized designs, to his career-long role in elevating the stature of the design profession.

Born in 1934, the youngest of four children, Danne grew up on a farm near Kingfisher, Oklahoma, during the height of the Dust Bowl. Early on, he learned the importance of hard work, doing much with little and sharing his talents for the benefit of all. In that vast open space he first called home, Danne became a top student and a versatile trumpet player. He had no art classes, but drew constantly. He developed an early confidence and open mind with encouragement from his family. Fortunately for the world of design, Danne did not follow his early jazz calling, after a summer spent on the road. It was a turning point, although the trumpet and a Steinway piano remain important to this day.

After attending Oklahoma State University and UCLA's Graduate School of Design, Danne set out for Dallas, where he immediately won clients and accolades, catapulting himself to the center of the local design community. Sensing there was more to discover, he and his wife, Barbara, took a leap of faith and moved to New York City in 1963, with a toddler in tow. He established himself quickly, becoming a partner at Gips & Danne in 1964, and later at Danne & Blackburn in 1973, although he has led his own firm, Richard Danne & Associates, now DanneDesign, for most of his career.

Danne's annual reports, created from the late 1970s to the early 1990s, defined an era. In a time when the annual report was a corporation's key branding statement, Danne designed books for such well-known companies as DuPont, Potlatch and Seagram, using gorgeous photography and elegant but simple typography to tell meaningful stories. He displayed a rare gift for convincing clients to trust a creative process that ultimately yielded beautiful, award-winning pieces. His longtime clients included household names – AT&T, Bristol-Myers, Pratt & Whitney, Bell Labs, NASA, the Fashion Institute of Technology, South Street Seaport Museum, Memorial Sloan-Kettering Cancer Center and Harvard Business School. He has won national and international awards, including three U.S. Presidential Awards for Design Excellence.

Over the years, Danne has devoted countless hours to his other passion: creating a community of graphic designers and enhancing the profession's standing. Elected to the Alliance Graphique Internationale in 1974, he was president of AGI/USA for many years. As president of AIGA from 1977–1979, he helped set in motion innovations that make the organization the leading voice of the profession it is today. He also organized and served as founding president of the New York Chapter of AIGA.

Danne leaves a lasting legacy through his work – elegant, understated, modern, beautiful.

An Update

As of late 2023, AIGA has 70 chapters and some 15,000 members.

AIGA/NY, the largest chapter, celebrated its 30th anniversary in 2012 and I was proud to participate in NYC events.

AIGA/SF, the second largest chapter celebrates its 40th anniversary in 2023. And because we now live in the Bay Area, I'm a member of this superb chapter which honored me with their 2018 Fellow Award.

For their 40th Anniversary, I've written a history of the San Francisco Chapter's birth and its pivitol role in AIGA's national expansion.

Since 1965, I've worked closely with five AIGA executive directors.

AIGA / SF 2018 Fellow award.

What is the secret of a 30-year consultancy?
It requires dedication and commitment,
from both client and designer, plus patience,
love, and a healthy sense of humor!

30 Years of

A One-of-a-kind College.

F.I.T. has been the best experience of my life.
Judith Pare '78

The classes are great and the city is fantastic.

The reputation of the teachers brought me to F.I.T.
Fred Santiago, '78

It's a place to develop a style of your own.
Donna Jantzen, 78

I wouldn't give up my place here for Harvard, Yale, or any institution in the world.
Clarence Babcock, 78

Excellent program, super teachers, very impressive.
Deirdre O'Brien, 78

Everyone says that the Placement Office is great.
Bernard Samuels, 78

New York is the best and only place for the college.
Laurie Troske, 78

This is one of our early recruiting brochures published in alternate years for high school students. It is tabloid-size with student quotes wraping around all covers. 1976

Nancy Yedlin was head of FIT College Relations Division which encompassed all communications, recruiting, and marketing for this unique college. It was 1969 and she had noted my award-winning work for State University of NY, of which FIT was a member college. Her inquiry was simple: "Would you like to tackle our next View Book?" (the trade jargon for recruiting brochure). Our interview was delightful and, frankly, it seemed pre-destined. For all the serious budget constraints over the next dozen years, we created imaginative work, won gold medals for Nancy, and thoroughly enjoyed the trip. In 1982, she wrote:

"...You and your staff have been a total delight to work with. Creative, responsive, patient and understanding, committed and striving for perfection, you always somehow manage to maintain your sense of humor. No wonder I want our relationship to be a continuous one."

Nancy contracted cancer a few years later, and she succumbed after a long and wrenching struggle. I lost a very dear friend and kindred spirit.

Two other spreads from the 1976 recruiting publication:
New York City as extended campus;
Quotes about the college.

As fate would have it, Loretta Lawrence Keane, Nancy's right arm and talented co-worker was being considered for her job. I liked her immensely, so I jumped in and campaigned for Loretta with President Feldman. She did get the job and performed brilliantly over many years in spite of some challenging times at the college (it seemed flux was the modus operandi). But, we all loved working together and continued to generate award winners. And when I finally resigned in 1999, our pal Loretta sent these thoughts:

...In fact, as I consider the significant change that has occurred here at the college during the several years, it is great testament that you and your work were often the stabilizing influences.

They say all good things must come to an end, and I guess this is our time. I will, however, count on your counsel and friendship for a long time to come."

It doesn't get any better than that –
one-of-a-kind college; one unprecedented run; two priceless friends.

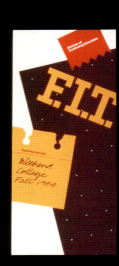

Continuing Education brochures
designed for two-year use,
with black text on interior pages.
1984–1985

F.I.T.
Pick of
the
Big Apple

A State University of New York college
for business and design professions

Fashion Institute of Technology
Seventh Avenue at 27 Street
New York City 10001-5992
212 760-7675

*"The FIT Story" published in
alternate years for high school
recruits. This poster / mailer
has text and photos of college
life on its backside. 1986*

F. I. T.

Fashion Institute of Technology

A College for Design and
Business Professions
State University of New York

L O O K

B O O K

A later edition of FIT's view book.
I titled it LookBook and the
college used the brand for decades.
Filmstrips of students and teachers
are both animated and fun. 1992

Advertising and Communications

Students in the Advertising and Communications major prepare for careers in advertising, publicity, public relations, publishing, journalism, broadcasting, and marketing communications. They learn to communicate ideas that influence people through a variety of media, including advertising campaigns, press releases, newspaper stories, commercials, and magazine articles. Students also write, produce, and direct television and radio commercials and multimedia presentations.

Many students gain on-the-job experience through externships with advertising agencies, public relations firms, newspapers, magazines, and television and film producers.

In addition, the program offers practical experience through the Advertising Club, the Public Relations Student Society of America, *W 27* (the student newspaper), *WIT* (student radio station), and the annual American Advertising Federation college competition.

As a graduate of the A.A.S. program, students may apply for admission to the upper-division majors in Marketing (Marketing Communications option) or Production Management. By taking certain course requirements, graduates also may apply to the Cosmetics, Direct Marketing, Home Furnishings, International Trade, or Merchandising Management options of the B.S. major in Marketing.

Fashion Buying and Merchandising

The foundation for a buying and merchandising career is built on a sound grasp of the practical aspects of this part of the industry. The curriculum includes courses in buying, merchandising, retail operations, fashion coordination, product development, advertising, direct response, product knowledge, styling, and sales. Potential merchandisers learn about garment construction, production, and costing in order to evaluate fashion merchandise.

The creative approach to merchandising and fashion direction is reinforced by visits to designer showrooms, wholesale firms, and retail buying offices. Students organize traditional runway shows and other fashion presentations. They may also gain cooperative work experience.

Qualified students may apply to spend a semester abroad studying merchandising.

F.I.T. graduates seek careers with retail, wholesale, and buying organizations. Graduates may apply for admission to each upper-division major in Fabric Styling, leading to the B.F.A. degree, or to the Bachelor of Science majors in Marketing (Cosmetics, Direct Marketing, Home Furnishings, International Trade, or Merchandising Management options) or Production Management. By taking certain course requirements, graduates also may apply to the Marketing Communication option of the Marketing major.

Display and Exhibit Design

More than walls, floors, props, mannequins, and signs, display and exhibit design is the successful relationship of ideas and objects in space. From shopping bags and store-wide promotional events to stage sets and gallery exhibitions, this design form is key merchandise, lighting, fixtures, graphics, and sound. Museums, galleries, showrooms, manufacturers, and trade shows all call upon the services of exhibit design experts.

Students in this major learn how to develop a total visual presentation. They execute small product showcases and large concept requiring construction. Majors acquire a knowledge of three-dimensional studio design, drafting and rendering, graphics and signage, point-of-purchase, visual merchandising, and exhibition practices. Students must also become familiar with fine arts, merchandising, photography, lighting, and packaging design.

A graduate is prepared as a visual presentation designer, display director in department or specialty store, or display or exhibition studio artist. Graduates also may apply to each upper-division majors in Restoration or Toy Design, leading to the Bachelor of Fine Arts degree.

Two spreads from the 1992 view
above: classes by major
left: the publication flips for the
happy "Student Life" section.

STUDENT LIFE

The Campus

FIT's eight building complex is home to more than 4,000 full-time and 5,000 part-time students.

Facing Seventh Avenue is the Shirley Goodman Resource Center, which houses the Library, the National Museum of Fashion, the Galleries at F.I.T., and the Graduate Studies Division. The Library serves everyone from those in equipment with materials used by both liberal arts and the technical aspects of an F.I.T. education. An international assortment of sketchbooks, periodicals, and business records is available, as are slides, tapes, films, extensive clipping files. The National Museum of Fashion holds the world's largest collection of costumes, textiles, and apparel; students, museums, scholars, and researchers make dramatic use of it. Two busy rooms hold talented collections of clothing and accessories, dating from the 16th century to the present, and friendly millions of textile swatches.

Recent exhibitions in the Galleries include *Elsa Peretti*, a retrospective of the sculptor's 20 years of jewelry design; *Seeing Yourself: Progressive Advertising in Television*, a retrospective of the great television commercials about young fashion; *A History of Eveningwear in the 20th century*; *Giorgio Armani: Images of Man*, a showcase of the photographer's way era commissioned by Giorgio Armani to document his work; clothing since 1970; *Fashion in Film*, an exhibition of costumes and photographs from current and historical movies; *Art Fashion: Important Fashion Masters Icons of the Twentieth Century*, a demonstration of fashion illustration in both a critical and creative art; *Pictures of Peace*, a thought-provoking collection of photography by prominent photographers; *Madame Matisse Retrospective*, an analytical study of Matisse as a designer of essential living; and *Tribute to The Black Fashion Museum*, an African-American contribution to fashion in celebration of The Black Fashion Museum in Harlem.

Facilities in the Art and Design Center include two fully furnished photography studios with one lit vehicle color darkroom and one black and white darkroom with 30 enlargers; three painting rooms; a sculpture studio; a printmaking room; a graphics laboratory; display and exhibit design rooms; life-sketching rooms; and a model-making workshop.

Also in this building is the Katie Murphy Amphitheatre, a striking

*The 50th Anniversary Mark
and palette of colors. 1993*

*Anniversary apparel designed for
sale in the FIT Student Shop. We
teamed with students from the
Merchandise Society to create these
and other products like mugs,
buttons, and tote bags. 1993*

F. I. T. FIFTIETH
ANNIVERSARY
EDITION OF
THE LOOKBOOK

50

Fashion Institute of Technology

A College of Art and Design,
Business and Technology
State University of New York

F. I. T.

A College of Art and Design,
Business and Technology
State University of New York

Network

F.I.T.

50

Making a difference

In 1944, FIT was
a great idea. In
1994, it was a pow-
erful reality mak-
ing a difference
each. FIT is the
American dream.

An oversized, spirited Newsletter for this unique college. "Network" editorial content reflects the hip, distinctive nature of staff, students and alumni. 1990

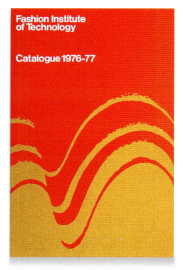

All-College Catalogue series designed for continuity. 1976–1981

F.I.T.

Spring 1995

Schedule of Classes
Division of Continuing Education
Fashion Institute of Technology
State University of New York

27th ST

Schedule of Continuing Education
Division of Continuing Education
Fashion Institute of Technology
State University of New York

Fall 1994

F.I.T.

F.I.T.

Summer 1994

Schedule of Classes
Division of Continuing Education
Fashion Institute of Technology
State University of New York

27th ST

Winterim 1995

F.I.T.

Schedule of Classes
Division of Continuing Education
Fashion Institute of Technology
State University of New York.

27th ST

*Ad series for a newly created
Toy Design Department.
There were no student-designed
toys as yet, so we created a
playful mood instead. 1993*

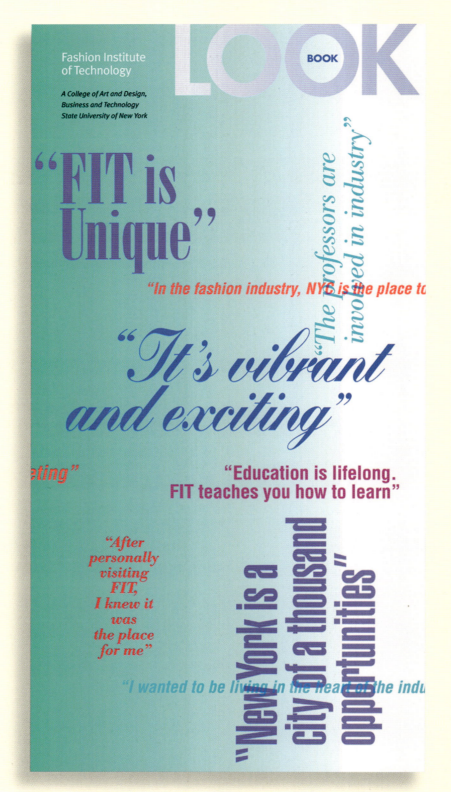

Fashion Institute
of Technology

A College of Art and Design,
Business and Technology
State University of New York

LOOK **BOOK**

"FIT is Unique"

"The professors are involved in industry"

"In the fashion industry, NYC is the place to

"It's vibrant and exciting"

eting"

"Education is lifelong. FIT teaches you how to learn"

"After personally visiting FIT, I knew it was the place for me"

"New York is a city of a thousand opportunities"

"I wanted to be living in the heart of the indu

*This was our 16th edition of the
LookBook over 30 years. An oversized
format with quotes on all covers. 1998*

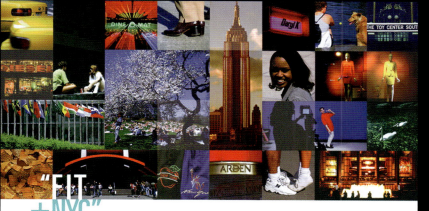

"FIT
+NYC"

What does it mean to students when they are told NYC and FIT merge as one? Why should they care? The meaning is significant because it goes to the heart of an FIT education. Just as New York City has set the standards for cosmopolitan life in the arts, business, communications, fashion, and culture, FIT has set the standard in education in art and design, business and technology. New York City's creativity and boldness have been duplicated at FIT in its vigor and excitement in classrooms, studios, computer facilities, and laboratories.

New York City has inspired FIT, a college of the State University of New York, to be a groundbreaker, setting the pace in career-driven education by taking the best the city has to offer and building on it. FIT has 30 degree programs, all developed in response to industry's needs. Many of them are innovative and one-of-a-kind, such as the country's only two-year program in Accessories Design and the only bachelor's degree programs in Home Products Development and Marketing, Direct Marketing, and Cosmetics and Fragrance Marketing (where students learn in the only fragrance develop-

ment environment on a college campus). FIT's educational offerings in Fashion Merchandising Management and Display and Exhibit Design were the prototype for others that followed. And, the college's Advertising and Marketing Communications program is the only one to teach communications from a marketing perspective. All of these innovations reflect industry's changing demands and FIT's uncanny ability to anticipate those needs.

Students might also ask why FIT, Fashion Institute of Technology, when so many programs offered are in fields that seem—at first—unrelated to the fashion industry. This answer also is at the heart of their education. FIT's name is a proud acknowledgment that the college has remained true to its original mission to educate individuals for work in the fashion industry. However, the industry has transformed itself dramatically over the decades, as has New York City, and FIT's leaders have had the foresight to expand its programs not only in fashion and its related fields, but into other areas of study where industry has made New York City its focal point and the consumer is key, such as Interior Design and Toy Design. Like the city and

From the same 1998 view book.
left: several spreads feature
New York City as ultimate campus
below: showcasing the great
latitude the FIT student will enjoy.

ACADEMIC FLEXIBILITY

Choices. People love to have them even if they sometimes confuse them. And FIT's worldwide reputation for curricular innovation has been earned in part by offering an elaborate portfolio of programs that complements an equally eclectic class of student. There are associate degree programs that can be completed in the traditional two years, and one-year associate degree programs for those who have already earned appropriate college credits. There are bachelor's and master's degree programs providing advanced study and the chance for further development of creativity and personal style. FIT's Presidential Scholars program offers honors students specially designed courses and colloquia. FIT students also have the opportunity to learn out in the real world — and all over the world — through one of the college's many internships and international study abroad programs. There is even a visiting students program that gives students from other colleges the opportunity to take courses without a formal transfer. With classes held morning, noon, and night, right around the calendar, students have the flexibility to meet their academic goals. It's choices like these that enable students to schedule their classes in ways that accommodate life's other demands. A little confusion in a flexible world is not a bad tradeoff.

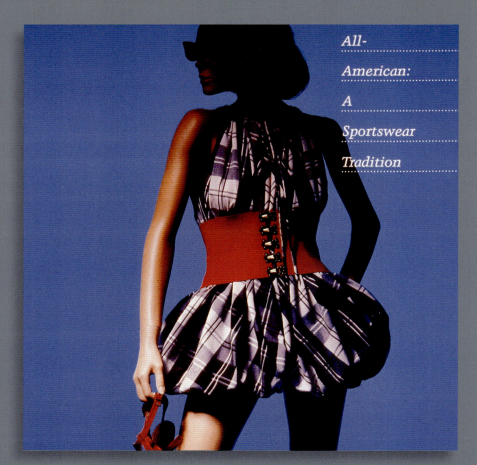

All-
American:
A
Sportswear
Tradition

Tip of the iceberg!

We have shown only a smattering of the FIT products designed over three decades. Also included were jillions of small space ads, specialty brochures, one-of-a-kind marketing pieces, seasonal promotions, and branding.

It is a testament to the individuals involved that we all enjoyed every minute of this special relationship in an industry that is built on, and demands "change!"

There were many special projects over the years. Here is the catalogue for a historical exhibition featuring American sportswear design, staged for The Museum at FIT. 1984

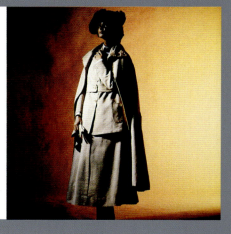

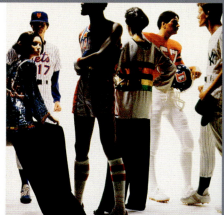

This project embodies the many responsibilities of today's visual designer. Many people don't realize what a professional designer does to make it all happen, but it's a lot.

World Flight

AMELIA EARHART ~ WORLD FLIGHT 1937

WORLD FLIGHT 1997 ~ LINDA FINCH

The dust jacket of our hard cover
book for Pratt & Whitney, the renowned
engine manufacturer.
P&W built the radial engines and sponsored
the historic flights of both
Amelia Earhart and Linda Finch,
exactly 60 years apart.

melia Earhart was a successful aviator – the darling of an era – and model to young girls all across America and around the world. Often referred to as an original feminist, she retained those special qualities that made her popular with both genders, popular beyond measure. Her tragic loss, attempting to circumnavigate the globe in 1937, was incalculable to the American psyche.

By the Fall of 1996, H. Christian Conover had been a client for some decades. Our relationship had spanned several organizations including the New York Power Authority, United Technologies, and Pratt & Whitney where he now worked. Chris called one day, "How would you like to do a book about Amelia Earhart's *World Flight?*"

"Are you kidding? Yes, of course!" I said tripping over my own enthusiasm.

Pratt & Whitney had built the original Wasp engines for her Lockheed Electra back in the 30s and been an important contributor to her success. And now, some 60 years later, a Texan aviator by the name of Linda Finch was planning to duplicate that *World Flight* voyage – but with a very different ending!

A lot had happened during the 60 years between these two iconic flights: A World War, technological breakthroughs like computers, trips to the Moon, radical social change, and significant style shifts – perhaps the greatest time of change the world had known to date.

When approaching the book's concept and design, I decided Amelia's section should be treated in a very personal way with mellow colors like sepia and cream, serif typography, borders and rules, and finished with matte varnishes – all very retro and comfortable. Conversely, Linda would be treated in a more modern way with crisp silver backgrounds, strong black & white images, sans serif typography, and glistening gloss varnish. The separate parts of the large book would have a transition sequence to fuse these two World Flights of such dramatically different eras.

Amelia Mary Earhart, 1903.

Amelia in Los Angeles, 1921.

The search for unusual, quality images took us to Radcliffe College, San Diego Aerospace Museum, United Press International in NYC, the National Air and Space Museum in D.C., the NY Times, AP/Wide World Photos, and most importantly, to Purdue University in Indiana. Amelia had a strong and enduring relationship with Purdue and intended to settle down there after her *World Flight* project.

Earhart was awarded the Distinguished Flying Cross by the U.S. Congress.

Amelia and her Lockheed Vega, before flying solo across the Atlantic, 1932.

With Edward Elliot, president of Purdue, who purchased Amelia's plane for her "World Flight 1936."

Her jack-in-the-box exit in Dakar, Senegal, 1937.

People often ask what a graphic designer does and I can answer, "Most everything!"

It's true, and this book project is a good example. We had conferences with the client throughout, developed fees and expense budgets, wrote schedules, hired the writer, selected the photographer, conducted all research for archival images, provided ongoing photo direction, edited all photos, designed both initial and final presentations, and coordinated the editorial text, wrote printing specs, solicited print bids and recommended the printer, prepared finished electronic art, camped out at the printer for a week of on-press supervision, specified custom shipping cartons, coordinated the book binding and delivery of all finished materials to the client in Hartford, Connecticut.

Everything above was contracted through our firm, so we also had the financial responsibility for the turn-key project. A cool $265,000 back in 1997!

Linda Finch, prior to "World Flight 1997."

It took over a year to find an Electra and another year to refurbish the aircraft. Here it is on its initial test flight in 1996.

Finch images by chase plane photographer Nancy Moran.

*Finch pilots the rebuilt
vintage Electra.*

So, was the project completed successfully? The answer is an emphatic yes!

May 28, 1997, and the end of a 60-year-long road. Linda Finch touched down in her Electra... she had finally finished the 26,000 mile flight that Amelia could not. The crowd cheered her arrival and, after the press conference and interviews, the plane was towed back to the exact same hanger that Amelia had used in 1937.

Acme, the great Boston printer, did a brilliant job on the books which were delivered several weeks ahead of schedule to P&W's many customers and VIPs around the world. This truly was a grand slam for the client and wide smiles radiated from Hartford.

But there is more good news. The *World Flight* book racked up some impressive awards that next year, however one was extra special – in a global printing exhibition, it was judged Best International Print Job of the Year.

Winging southeast from Indonesia to Darwin
over the placid Sea of Timor.

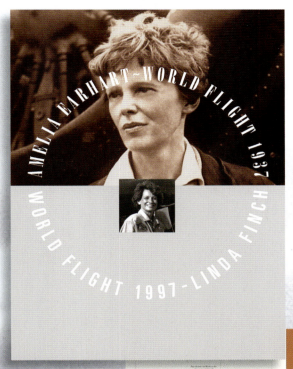

AMELIA EARHART ~ WORLD FLIGHT 1937

WORLD FLIGHT 1997 ~ LINDA FINCH

*The finished hard cover book,
with copper and silver dust jacket. 1997*

Second Attempt

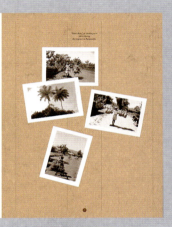

"Amelia Earhart came
perhaps before her time."
WALTER J. BOYNE
Aviation Historian

Always keep your eyes open for ways to maximize the assets in front of you. Sometimes great possibilities are hiding in plain sight.

Group W Albums

Sometimes the best concepts and ideas are so obvious you might not recognize them and miss a golden opportunity.

Group W, the newly minted marketing moniker for Westinghouse Broadcasting, had been producing radio documentaries in the early 1960s; radio was still a very powerful medium across America, and these were great, award-winning programs worthy of note.

The youthful and savvy Tad Ware was the Communication director charged with projecting a strong image for the Group W enterprise. Gips & Danne was also a new partnership which was attracting attention in New York's design world. Those documentary broadcast tapes were just sitting on the shelf so, collectively, we hit on the idea of using these astutely researched audio tracks, repackaging them with striking graphics as LP albums, and distributing to the most important movers and shakers in the industry. As a gift.

It was a brilliant idea. The content was absolutely free and deserved a larger appreciative audience. To our surprise, the production costs for this albums was quite low. And our very first album was a hit – the recipients were ecstatic!

This was a promotional gift that couldn't be thrown away! The valuable content was presented in such a unique and compelling way that it had to be kept.

"The Music Goes Round and Round"
A five LP set features pop music
from Tin Pan Alley to early Rock &
Roll. Label font changes for genre of
music. The booklet inside describes
the scope of this documentary. 1967

Winston Churchill was indeed "The Lion's Heart" and this single LP came in a jacket with memorable graphics. 1965

THE GREAT ONES Famous American Negroes and their achievements. Based on a series of radio programs broadcast by Group W Westinghouse Broadcasting Company, Inc.

"The Great Ones" showcased notable black Americans and their profound achievements. The five LP set has custom labels for each person. 1966

Milestone days of World War II are memorialized in this three LP set. The box container is lined in red to commemorate the sacrifices made on these momentous days. Collage is built from newspaper headlines of the days. 1967

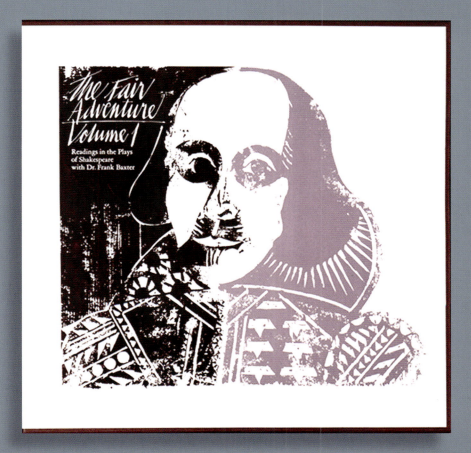

"The Fair Adventure" is a two-box series
exploring the major works of Shakespeare.
Multiple records in each box feature
custom labels with quill-like handwriting.
Illustration by Seymour Chwast. 1968

"The Fair Adventure" Volume 2 is a companion to the previous and features a different group of Shakespeare's plays and writings. 1968

It's important to engage in larger issues and causes for the public good. One of our long-term interests is the environment, something we all can care about.

Green Stories

People can differ on why the world's climate is changing so radically, but there is little argument that it is. Temperatures rising, Arctic ice thaws, higher sea levels, marine migration altered, social disruption, more frequent and severe hurricanes, fires, and storms. Climactic weather!

Our firm's client roster always included one half non-profit organizations. We felt it was important – an obligation to support the public good. This section features two long-term clients whose timely, critical narratives yielded successful fund-raising campaigns and program goals achieved:

Woods Hole Research Center

Center for Coastal Studies

Forests of the World

The Woods Hole
Research Center

Most people know that two thirds of the world is made up of water, but not many realize that nearly half of the land surface was once forested. Since the advent of agriculture about ten thousand years ago, humans have burned, cleared, chopped and sawed forestland for fuel, farms, a place to live, and timber for construction. In much of the world, forests have been replaced by an impoverished and even barren landscape.

Forests clean our air, moderate our climate, provide for steady supplies of clean water and a wide range of resources from timber to food. They provide habitat for indigenous people and for millions of plant and animal species. Forests, including their soils, hold more than twice as much carbon as the earth's atmosphere.

The loss of forests degrades the landscape, affecting not only local communities, but also people everywhere. Their destruction adds carbon dioxide to the atmosphere resulting in a rise in the earth's temperature, the disruption of climate, and a rise in sea-level. And scientists believe that nearly half of the world's species of plants, animals, and microorganisms will be destroyed or severely threatened over the next quarter century due to rainforest deforestation.

AMAZON

AFRICA

RUSSIA

ALASKA

Where We Work

Though headquartered in Woods Hole, Massachusetts, the staff of the Woods Hole Research Center focuses on five major forested areas of the world: the tropical forests of the Amazon Basin, the forests of Central Africa, the boreal forests of Eurasian Russia, and the forests of North America, especially in New England and Alaska. Each of these regions has an immediate need for basic research in ecology and for the development of government policies that support and enhance conservation.

What We Do

The Woods Hole Research Center produces the fundamental scientific understanding of the global environment and promotes conservation worldwide.

In the Amazon Basin, we conduct ecological and economic research aimed at improving public policies designed to conserve the world's largest rainforest. Our field experiments simulate forest responses to a drier climate and a period of intensified fires. Our studies of the timber, ranching and soy industries identify ways of reconciling economic growth with forest conservation. We have launched regional planning processes along the giant highways that are slated for paving into the heart of the Amazon, where development will almost certainly cause thousands of hectares of deforestation along the newly traveled corridors.

Continued on reverse side

The **Woods Hole Research Center*** focuses on issues of global climate change and preservation of the world's great forests.

WHRC was founded by renowned ecologist George Woodwell, a collaborator with Rachael Carson of "Silent Spring" fame. It's an organization of research scientists dedicated to finding answers to the great and disruptive climate issues of our times. With an administrative staff at their "green" headquarters in Woods Hole, MA, the Center's personnel are now deployed all around the world studying thawing permafrost in the Arctic to deforestation from Indonesia to Brazil.

Growing up on an Oklahoma farm in the 1930s, I became acutely aware that the weather directly affected whether we lived poorly or well. My Dad and our family learned to respect nature and deal with its power and consequences. Small wonder that decades later I would be drawn to the people and the work of the WHRC.

I met Dennis Dinan, director of Development, soon after coming to Cape Cod in 1995. He was quite skeptical at first, about hiring a "New York consultant" for this important mission. But what ensued was a great working relationship that lasted over a decade. The content was meaningful, and we got the desired results with one successful fund raising campaign after another. I loved the work!

And when we left the Cape for Napa in 2005, Dennis delivered this evaluation: "The firm has created superior products that, truly, are the envy of our peers."

"THE SEASONS" A campaign series of mailers employing Native American sayings about protecting and preserving the Earth for future generations. 2002

Opposite: Fund raising poster / mailer featuring WHRC scientists studying forest preservation around the globe. 2004

*Now renamed:
Woodwell Climate Research Center

*Fund raising poster / mailer featuring
the Build Out development of the
Cape over a 40-year period. 2003*

Kilaparti Ramakrishna is uncomfortable with superlatives. He has a lawyer's refined instinct that the way to accomplishment lies somewhere between extremes, and words like biggest, most, and least are not likely to appear in his writings or conversation. As a negotiator at the forefront of the environmental debate among the nations of the world, he avoids polarities as carefully as he would shun an argument made from a platform of weak science.

So it is an irony that, when others describe the deputy director of The Woods Hole Research Center, *best* is the word

As he stepped into a taxi recently at the White House, Professor John Holdren was asked by the driver what he had been doing *in there*.

Holdren explained that he serves on the President's Committee on Science and Technology and that he had been *in there* to advise the President, particularly on matters concerning the environment and arms control.

"Good," said the cab driver, "you can give him some advice from me." There followed some simple, practical suggestions of a kind that Holdren believes the

In the spring of 1950, as the North Koreans prepared to make war against their brothers to the south, and in Moscow their Soviet sponsors were forcing entry into what would be called, ironically, the Nuclear Club, George Woodwell, a senior at Dartmouth College, won the Eggleston Prize for exploring the limits of a local bog called, without irony, the Bottomless Pit.

The Korean War went on for three years, the nuclear arms race forty-five, and George Woodwell ever since has been trying to pull us out of one bottomless pit after another. As he proved at

"VOICES" A brochure series profiling the prime scientific leaders of the Center. Our first assignment for this client. 1997

Annual Report series, each featuring a different theme: Tree species, Scientific staff, Headquarters new wing, and Global activities. 2001-2004

CELEBRATING
A MILESTONE

Our scientists and policy staff
are dedicated to solving the complex
issue of climate change and
to defending Earth's great forests.

The Woods Hole Research Center
offers a model for a new era
with its Gilman Ordway Campus,
dedicated on June 7, 2003

The commemorative poster / mailer for
the opening of the new WHRC
science and research facility. 2003

"VISION" A tabloid-size newsletter to track Capital Campaign progress for a new "green" WHRC headquarters, Woods Hole, MA. 2001–2002

As part of the Capital Campaign, a ladder-fold mailer with endorsements from notable leaders in literature, commerce, and science. 2004

*The Center for Coastal Studies
is well known for whale research and rescue – important
for the survival of these endangered mammals.*

Center for Coastal Studies

CCS 25

A milestone publication for the Center's
25th Anniversary, complete with historical
background and a comprehensive review
of current activities. 2000

Where all things converge

The Center for Coastal Studies sits at the end of the land and the beginning of the sea. Our vantage point, at the tip of Cape Cod, allows us to work in the midst of an extraordinary ecosystem where all things converge. It is the place where the temperate zone and the sub-boreal zone meet, where the sandy shore knows its northernmost reach and the rocky shore is about to begin. The explosion of upwelling nutrients just offshore and the convergence of many habitats means an explosion of wildlife as well, a home to feed and breed for shorebirds and turtles, whales and dolphins, seals and groundfish, sponges and periwinkles, sea urchins and sand dollars.

From our field station in Provincetown on this narrow, sandy spit of land, we can easily voyage into the greater Gulf of Maine to work on issues of habitat protection, ecosystem management, marine mammal and marine wildlife conservation. We fly aerial surveys over the Bay of Fundy; train fishermen along the coast of Maine; survey the humpback feeding grounds of Jeffreys Ledge off New Hampshire and Stellwagen Bank off Massachusetts; huddle over countless tables with shipping companies, municipal leaders, fishermen, and government officials to come up with new ways of solving old problems.

Our work carries us to collaborations throughout the Gulf of Maine, with such institutions as the U.S. Coast Guard, McMaster University, the New England Aquarium, the Woods Hole Oceanographic Institution, the Massachusetts Division of Marine Fisheries, the National Marine Fisheries Service, the Canadian Department of Fisheries and Oceans, the Maine Department of Natural Resources, and the University of Massachusetts. We welcome them back to our Provincetown port as well, where scientists and managers, professors and policy makers can use our facilities in pursuit of our common goal: the preservation of the sea, and its world around us.

*"Scientists have quietly begun a
whole new age of underwater discovery
that will be as challenging and as
exciting a mission as any we have
undertaken on land or in space."*
EXECUTIVE DIRECTOR PETER BORRELLI

The Center for Coastal Studies   This is a fine relationship which commenced on Cape Cod in 1997 and continued on for many years.

The Center focuses on marine mammal research and whale rescue around the world and on coastal and marine ecology and geology concentrating mainly on the Cape. From their base Marine Lab and headquarters in Provincetown, MA, the organization does remarkable work in the Gulf of Maine and other important ocean areas of the planet.

The scientific staff can be found in far-flung locations such as Tonga, Hawaii, American Samoa, Australia, Vietnam, Dominican Republic, Canada, and Chile.

My initial role was one of design and marketing specialist for all fund-raising projects of the Center. But somewhere along the way, I was invited on the board and into a more intimate involvement with the organization's leadership. I even ended up as board Vice-Chair for a period and enjoyed the work immensely.

When we moved west in 2005, I shifted to the CCS Advisory Board. I believe this is a valuable role for designers, not to just apply their professional chops, but to become immersed in the organization's structure and path and contribute to its very purpose and mission.

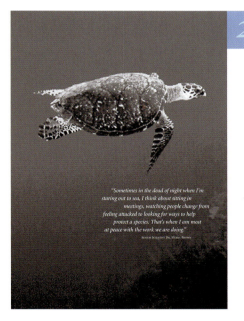

Interior spreads explore the Center's world-wide activities in marine mammal research and coastal ecology.

Presentation folder with inserts for giving opportunities. 2001

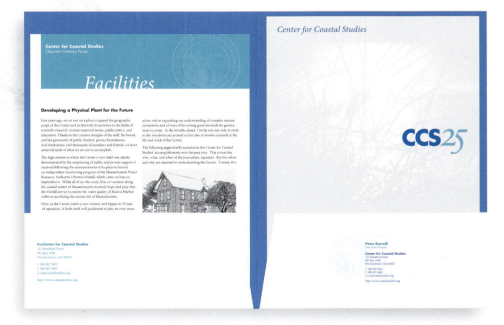

The commemorative, framed poster gift for major donors to the anniversary campaign. Painting by noted Province-town artist Paul Resika. 2000

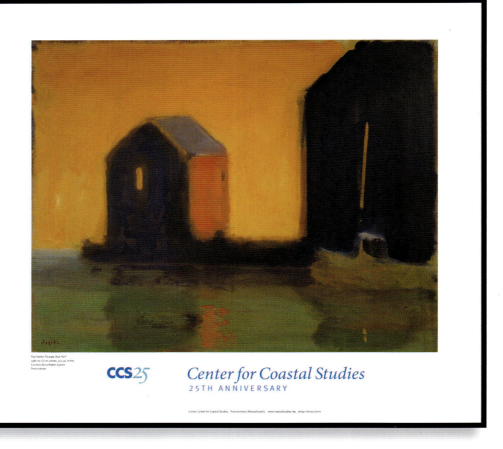

Invitation and envelope for a special 25th Campaign event staged at the striking JFK Library in Boston. 2002

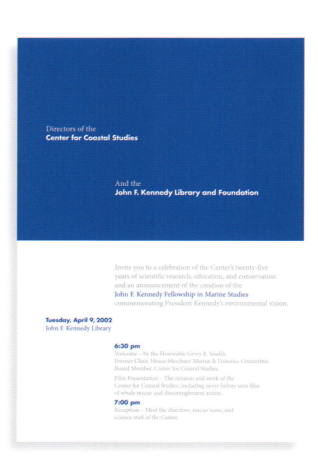

**Provincetown
Center for Coastal Studies**

COASTAL

*Brochure for a major campaign
to build a new Marine Science Lab
in Provincetown. 2004*

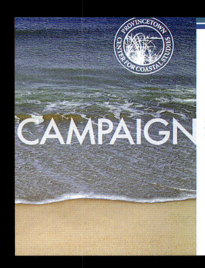

CAMPAIGN

Goal: $3.5 Million

*For Science and Protection
of Our
Marine Environment*

GOALS

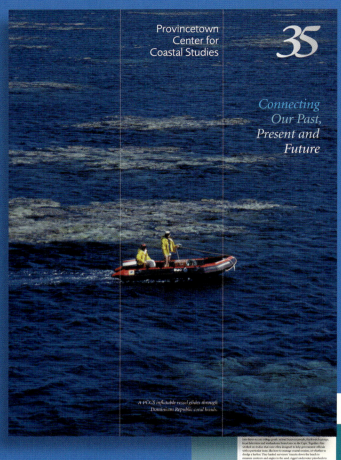

Provincetown
Center for
Coastal Studies

35

*Connecting
Our Past,
Present and
Future*

*A PCCS inflatable vessel glides through
Dominican Republic coral beds.*

*For the 35th anniversary, we created an
eclectic, informative publication with
editorial features and personnel profiles
from around the globe. 2011*

More leadership and service, but this time on the international stage. It was important to strengthen bridges with the global community.

AGI

Alliance Graphique Internationale

Prestigious
Elitist
Educational
Relevant
Argumentative
Friendly
Inspirational

The process of giving birth to the New York Chapter of AIGA was an ordeal! I was tired and this would have been a great time to take a nice long nap, a break from all this institutional work. Alas, that didn't happen; as I had already accepted the invitation to follow one of my true heroes, Gene Federico, and be the next president of AGI/US, the American contingent. He knew my history, that I got things done, and Gene implored me to take on the role. I just couldn't say no.

But don't feel sorry for me, this is a great honorary organization. An elite group of about 250 members, it was rather small at the time. And it represented a golden opportunity in many ways. The pressure was nothing like my previous ventures, it was a chance to work with my American member friends, and become closer to many international members I had long admired – the best of the best!

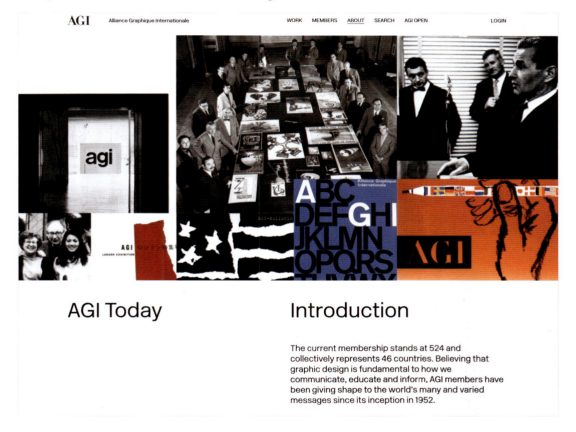

The AGI website, 2023,
posting 524 members from 46 countries.

First, here is a very brief history of how AGI began – taken from AGI's website:

"In the 1940s, commercial artists, mural makers, typographers, printmakers, art directors, illustrators and poster designers increasingly realized their common bonds, and the modern profession of graphic design began to be defined. In 1951, five graphic artists – two Swiss and three French – decided to formalize their relationship into some sort of association. Their idea was simply to share common interests and friendships across national and cultural borders. It was a notion that soon attracted leading exponents of the graphic arts from elsewhere in Europe and in the USA. In 1952 the Alliance Graphique Internationale was incorporated in Paris with 65 members from 10 countries.

Jean Picart Le Doux became the president, Fritz Bühler and FHK Henrion were vice-presidents, Jean Colin became the secretary general, and Jacques Nathan Garamond was the treasurer.

The first AGI exhibition was held in Paris in 1955 and in 1969 the headquarters moved from Paris to Zurich. Student seminars were introduced in 1979 and the first Young Professional AGI Congress was held in London in 1994."

I Am sterdam was the theme for the last Congress staged in The Netherlands, I created this animated short video using photographs taken on earlier visits. 2012

My poster contribution to an invitational exhibition produced by the Mexican delegation for the 100th anniversary of "Diego Rivera and Frida Kahlo." 2008

I was thrilled to be elected in 1974, as one of the youngest members at the time. AGI was much smaller and more elitist back then. But it was a significant honor, an important credential which I appreciate to this day. When I later served as AGI/US president for many years, it was a privilege. The role was fairly straight-forward, making sure the American members all got the current information in a timely manner. But because there was no internet as yet (thus no website or email) most communication was delivered by "snail mail" or by phone for more immediate matters.

Each year, new candidates from various countries are put forth for membership consideration. This task does require work as the coordination of these nominations is not uncomplicated, the nominee is required to have two sponsors plus visuals and biographical material. I've found designers to be among the most generous people on earth – I've also found them to be overloaded and consistently late with almost any kind of submission requirement. This is a universal trait and we all laugh about it, just part of our reality. The nomination of an AGI candidate can be like pulling teeth to get things done on time.

A massive compilation of the history of AGI, created by founder FHK Henrion in 1989.

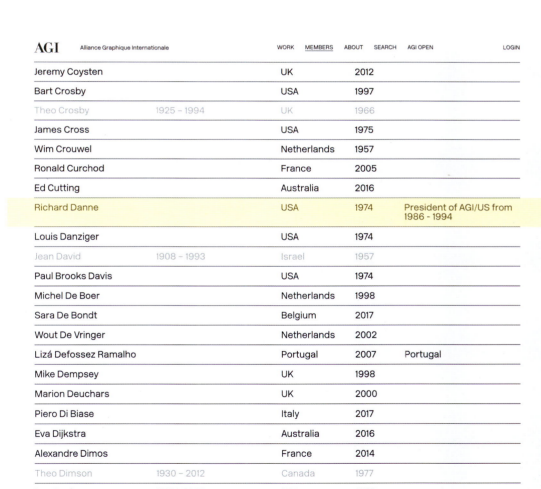

From the "Members" section of the AGI website. 2018

Every congress also has a design assignment including projects of great diversity, content, and medium. In some form, these assignments are showcased at the event, and they are wonderful and often inspirational. But, yet another item with a deadline! I'm sure you get my meaning.

However, and this is the best part, AGI stages its yearly congress in some wonderful or exotic place. However, and this is the best part, AGI stages its yearly congress in some wonderful or exotic place where one might never get to go under normal circumstances. For many of us, this means firsthand exposure to completely different cultures (i.e.: The Beijing Congress was staged before China was truly open to foreign tourist and visitors.) Often these congresses include government sponsorship and events, like the hospitality of the Mexican government during our congress in Oaxaca with a Governor's formal dinner and cultural entertainment.

And perhaps most important, these international members are able to meet, learn from each other, socialize, and develop lasting friendships with the greatest design talent on the planet. This concept of fellowship still drives the organization. In more recent years, a pre-congress educational event called "The Open" – focused on students – has become a staple and valuable part of each congress.

AGI São Paulo 2014

My poster celebrating the essence of Brazil for the San Paulo Congress. 2012

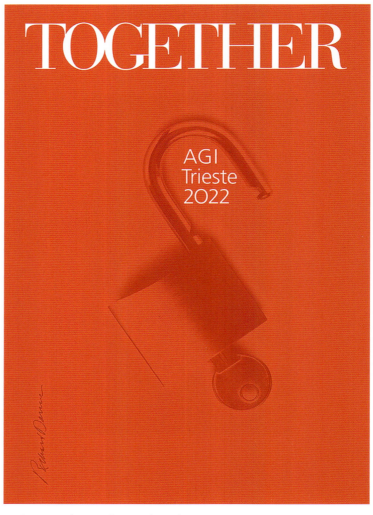

A Congress theme of "Together" for the Trieste, Italy, event after a three year hiatus due to Covid. 2022

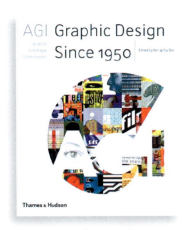

A handsome historical publication created by Elly and Ben Bos in 2005.

My multi-year tenure as AGI president culminated with the Montauk congress on Long Island, NY in May, 1993. Barbara, my spouse and professional partner, and I organized and produced this successful event. Wow, that's all I can say! What an incredible amount of work, but we did survive the congress – one of the larger ones. Our internal time sheets, back then, showed more than $200,000 in lost time, a likely average for anyone who produces one of these AGI events.

We were meeting by the Atlantic, at the easternmost point in the United States, so our congress theme was: "The Ocean." The theme inspiration was courtesy of B. Martin Pedersen, a member and Montauk resident himself. The congress turned out to be very well attended and was a hit!

One little side story: our program had many unusual speakers featuring subjects like underwater photography, international aquarium design, and many other "sea" topics, all highly enlightening and entertaining. But I also wanted to surprise the members with at least one unexpected, dramatic speaker and decided that Paul Rand was the perfect person. He had all the international respect, but he was notorious for not appearing in public – anyplace or anytime.

Members were challenged to design new surfaces for this iconic Finnish chair. Here is my submission at the Helsinki Congress. 2003

I negotiated with Rand for more than a month trying to convince him to come to the Congress and actually deliver his words of wisdom to this eminent assembly of celebrated designers. What it finally took was to wave all event fees (he was a longtime AGI member – and arrange a private car to escort Paul and his wife Marion Swanee, to and from their Connecticut home, which was a very long drive. There were other perks as well.

Paul was such a recluse, and we weren't sure until the couple exited that limo in Montauk that he would actually show up. Yet, he came out of his shell and delivered to an enthralled crowd, who couldn't quite believe this diminutive design giant was there and speaking to us.

The lesson? Sometimes, it takes an extraordinary effort to get the desired result.

And this Oklahoma farm boy can still marvel at a career that allowed him to be connected to and become friends with many of the world's finest artists and designers. That alone brings real contentment.

Barbara Danne mimics a sculpture at the Kroller-Muller Museum during a marvelous Congress in Holland. 1986

Dutch designer Ben Bos was among our favorite people. He helped produce that Amsterdam Congress, and this handsome little glass was a takeaway gift. Ben may be gone, but never forgotten.

AGI design assignment:
The rug under our Steinway piano was woven by local artisans in Oaxaca, Mexico before the congress. After being exhibited there first, the rug was delivered back to its designer. 1999

Here is a self-initiated project for the
National Air and Space Museum which was
successful and won major awards.

"Columbia" Chronicles

had been immersed in the U.S. Space Program for several years, when I decided to chronicle the "Maiden Flight of Space Shuttle Columbia" in a poster series. The goal was retail sales at the great National Air and Space Museum in Washington, D.C.

These posters featured strong, memorable images by Rene Burri, the famous Swiss photographer who was commissioned to track the original trek of our first U.S. Space Shuttle.

In titling the posters, I chose to give Columbia a persona – it helped the viewer identify with both spacecraft and its eventual crew of astronauts. The titles brought immediacy and action to the series.

NASA was not involved with the creation of these posters. They were beautifully printed by Hennegan, the great Cincinnati printer, who partnered with us on the "Columbia" project. Sold for many years at the NASM Museum Store, both as individual posters and as sets, the posters had a long and successful run.

A sad fact: Columbia suffered a catastrophic ending on February 1, 2003, when it exploded on re-entry over Texas with the loss of both spacecraft and all seven astronauts. It was a tragedy for the country, for the American Space Program, as well as myself.

Yet, the "Maiden Flight of Columbia" series lived on and was widely celebrated in prestigious publications and exhibitions.

First of the six 24" x 36" posters. The series
garnered many awards including a third
U.S. Presidential Award for Design Excellence.

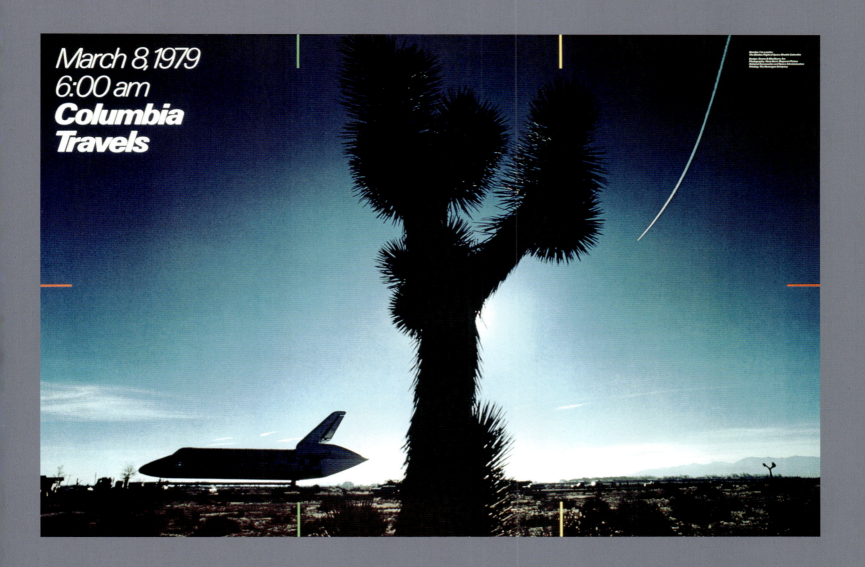

March 8,1979
6:00 am
**Columbia
Travels**

Crossing the vast Mojave desert, near
Edwards Air Force base in California.

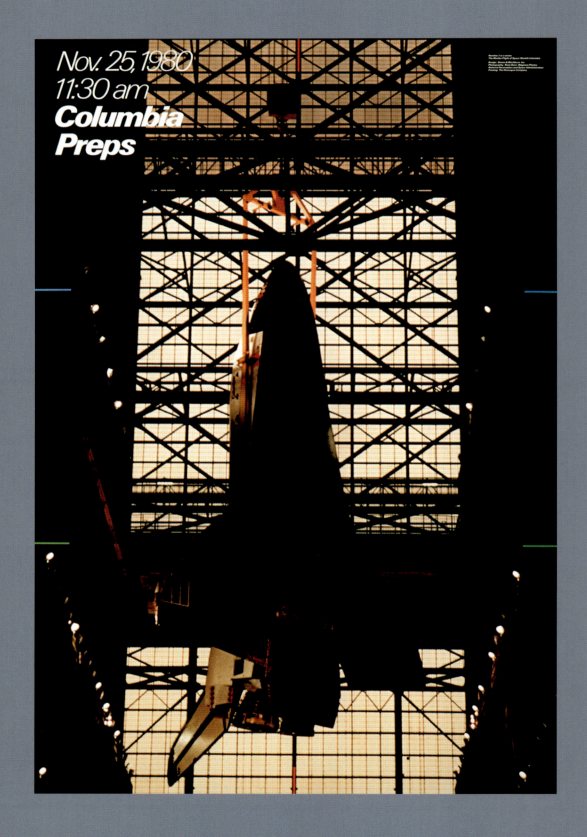

A very slow rollout as the Space Shuttle moves towards its Launch pad destination.

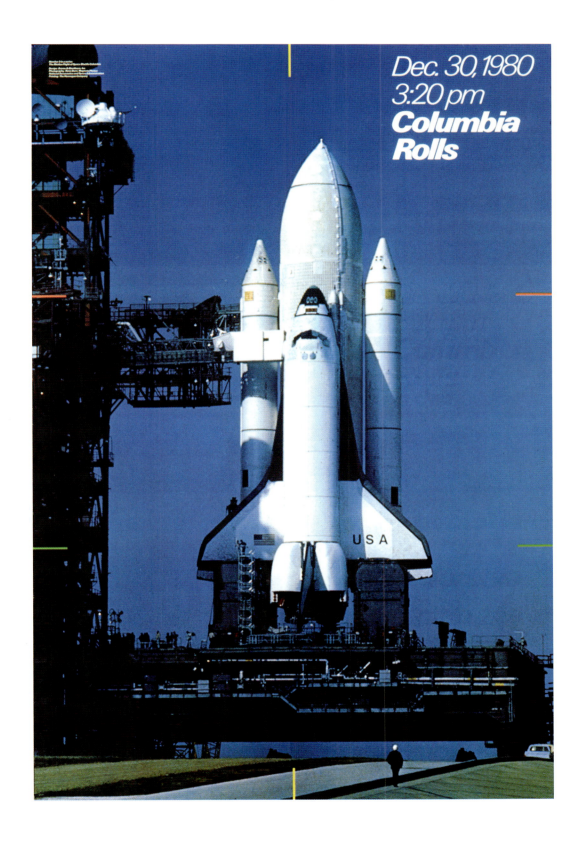

Dec. 30, 1980
3:20 pm
Columbia Rolls

April 9, 1981
7:50 pm
**Columbia
Waits**

*The Columbia trek spans two years
of time. This title refers to a
"hold" of several days on the launch
pad at Cape Canaveral, Florida.*

Our first Space Shuttle is airborne,
ready to orbit the globe on
its very successful maiden flight.

April 12, 1981
7:04 am
**Columbia
Flies**

April 14, 1981
10:15 am
**Columbia
Returns**

The Columbia series, created in 1981, continued to sell at NASM's retail shop for decades.

Music makes everything better!
And, here are three professional involvements
which have kept the music flowing.

Music + Design

Poster featuring musical compositions which I performed in high school competitions, 1951-52. The Haydn Concerto is one of the most respected in the trumpet repertory. The other three works are for piano. 2010.

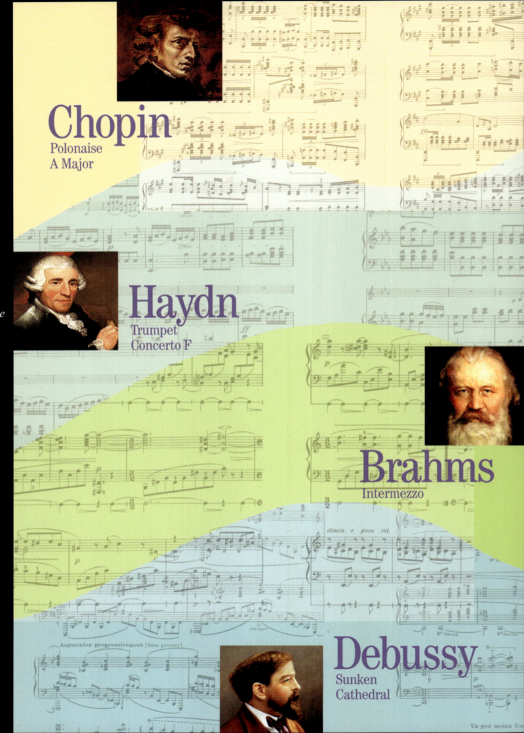

Music played an important role in our Oklahoma home. It brought much joy and was an uplifting element in our rather spartan family life on the farm. My siblings and I all studied and played the piano, though we had an old, no-name upright. My father was self-taught and played violin and guitar while Mother was an accomplished pianist who often accompanied us in competitions. Brother Art was a terrific alto saxophonist, and when he gave me my first trumpet, he gave me wings! I played professionally through my college years and came within a whisker of pursuing a career in modern jazz. Ultimately, design won the day, yet I've enjoyed the fusion of music and design over a lifetime.

Views of our beloved 1915 Steinway, Model M. This piano was built during the "golden age" of Steinway, and refurbished by us in New York, 1985.

Our modern jazz quartet, The Varsity Crew,
performs at the Cliff House, Colorado,
summer of 1955, with me on trumpet.
Painting by noted artist Hobart Hayes, a
regular patron and fan of the group.

My CD recording at age 88 with
improvisations on some originals and
favorite jazz standards. 2022

Third Street Music School Settlement   One of the great New York stories: "Third Street" was founded in 1894 by Emily Wagner as a true settlement school for children. That meant teaching young people how to play and perform on instruments but also to feed them, if necessary. And, in those earlier times, the need was great for both.

This singular school celebrated their 125th anniversary in 2020, as the oldest settlement school in the United States. I met the Music Director, Harris Danziger, in 1969, and began developing the rudiments of a pro-bono communication program. It was fulfilling to design a symbol for this wonderful school on the Lower East Side of Manhattan (which shared the block with the notorious Hell's Angels!) At that time, there were about 350 music students.

Around 1973, Third Street moved to Eleventh Street and to a significantly larger building in a better neighborhood. Refurbishing took almost two years but the school was now primed for rapid and dramatic growth.

In two different stints, I served on the board for a total of 18 years. Today, the school teaches some 5,000 students a year, in their own building and in the Public Schools of NYC. What a great school, and an unbelievable success story!

"Harmony On Third Street" was our first view book and it featured the Orff and Kodaly methods of teaching. 1969

The symbol is vibrant and still used for each and every marketing effort by the school – 50 years and counting! 1970

Posters announcing the School's move to
11th Street at St. Mark's Square, on the
Lower East Side. With the larger building,
the curriculum grew to includes studies
in full string orchestra, chamber groups,
jazz, rock, vocal, dance, and for all
individual instruments. 1973

right: this successful Capital Campaign
brochure was titled "An Investment in
Children and the Arts." 1986

*"A Century of Service" brochure for
Third Street's 100th anniversary. It features
a translucent cover with archival images
blended with contemporary photos. Also
shown are other anniversary products. 1994*

Cape Symphony Orchestra   There is nothing quite like a large symphonic orchestra performing at full force – it is a truly dynamic experience.

The CSO is one of the best regional orchestras in the country and we were invited to create a new design and marketing program for the organization. It all started with a new symbol to establish that a major change was underway. Then in logical succession, we initiated new advertising, marketing, and finally, their first web site. The program continued to develop over a nine-year period and was both cohesive and successful.

Enough so that the Orchestra moved in 1999 to a much larger, more effective performance hall with greatly expanded possibilities. Our main contact through those transformative years was Marketing Director Lisa Sheehy. Working together was delightful, and we built a solid professional and personal relationship.

As we headed west in 2005, she expressed these comments, "They handled all identity, marketing and advertising, and their designs produced tangible results in both ticket sales and increased Orchestra recognition."

It was a labor of love!

An ad series created when the orchestra was moving to a new venue. Goals were to announce the big move, and humanize the performers by highlighting their hobbies and avocations. 1999

CSO symbol anchors the entire program and embodies music notations, violins, and the sea. 1996

right: "Sounds for All Seasons" performance calendar with both classical and pops concert highlights. 2003

CSO's very first web site was interactive including ticket sales. 2004

The "Silver Season"
*celebrating the 25th Anniversary of
Conductor Royston Nash. 2004 - 05*

*Coordinated Program covers,
printed in Silver, for the full lineup of
Classical and Pops performances.*

*This rollout brochure was also
printed in silver and featured all of the
concerts for this milestone season.*

Napa Valley Jazz Society   Founded in 2010, the NVJS produces live concerts at various venues in the Valley. These shows have featured some of the best national and, occasionally, international jazz performers. These include top vocalists as well as instrumentalists for a nicely varied menu of talent.

As part of their educational mission, the Society awards scholarships to worthy music students and other grants to jazz programs at Napa Valley high schools.

Board members produce and promote their concerts which average 10 shows a year. I joined the NVJS board in 2012 and have enjoyed being involved in live performance again, even if I'm not on stage this time around. I've also been responsible for creating the group's website along with creating e-blasts models.

My music background has been quite valuable while working for Third Street, CSO, and NVJS. Everything comes full circle.

Home page of the NVJS website
with 12 of my live performance photos
in rotation. 2012
JAZZ logo by Jim Cross

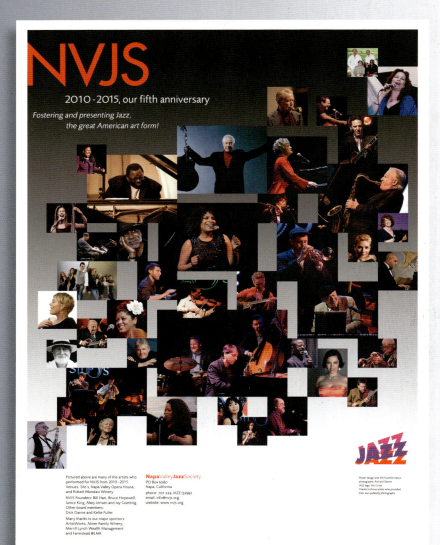

Poster to celebrate the Fifth anniversary of NVJS featuring all the performers. 2015

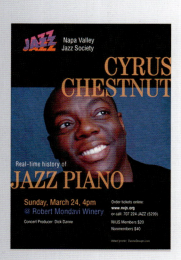

e-Blast series promoting individual concerts at different venues. 2014

Cross-cultural design has become more
important in recent years. And while
it can be difficult, it is incredibly satisfying
too as this M-J project demonstrates.

Branding in Br

Mr. J. Murillo Valle Mendes, president of Mendes Júnior, had just arrived at our offices on lower Fifth Avenue in NYC. He was accompanied by two associates, Pedro Albuquerque and Henrique Poeck – all from Belo Horizonte, Brazil.

They were shopping around New York in search of a firm to produce a strong identity program for their noted company, which had enjoyed a world-wide reach. Mendes Júnior Group was known for super-scale construction such as large bridges, subways, offshore rigs, and buildings of great variety and had worked around the globe – including Iraq after the first Gulf War. They were heavily involved in steel production too, with plans to become a major player in that arena.

They had already made the circuit in NYC with presentations by the biggest and best design firms in the City. They had scheduled weeks ahead of time and we were the final stop on their tour. We closed our eight foot conference room doors and launched into our show & tell featuring programs for some of America's best organizations.

Frankly, while we intended to give it our all, there were low expectations that we would win this competition. We weren't as flashy as those other design firms and we assumed they had brought out their heavy artillery. However, things went well with our presentation and they seemed quite alert and interested, not at all beat up from their fast-paced tour around New York.

View of Rio de Janeiro

We finished the presentation and Mr. Mendes made an unexpected request, "Would it be acceptable if our team might meet privately, behind closed doors?" I responded, "But of course, take as much time as you like." We had assumed that they would simply get on a plane back to Brazil and be in contact at a later date. But there was something different about this client – Mr. Mendes was the current CEO, but the company had been closely held by his family for generations. I would learn later that their employees were treated better than other workers in Brazil, with loyalty flowing in both directions.

About 30 minutes later, the doors opened and they asked me to rejoin their meeting. And to my considerable surprise, Mr. Mendes said, "We have selected your firm for the identity assignment, if you want to do it."
"Yes, we most definitely want to do it!" I replied. Having paid plenty of dues in ordinary places, a choice assignment in an intriguing country like Brazil... that's as good it gets.

I wondered aloud how they could make that decision so quickly (most programs can be weeks in suspension while deciding). Mr. Mendes clarified that for me: "You obviously have the talent and, just as importantly, you seem like a trustworthy firm." So, perhaps the lack of flash was working for us here?

The three gentlemen departed to catch their flight back home, and we agreed to immediately start the Phase 1 research process back in Belo Horizonte. They would set up the tour of facilities and appropriate interviews, and I would travel there within a couple of weeks.

Flights to Brazil were not plentiful, and the trip was long: 18 hours in total, including a flight to San Paulo or Rio, then another to Belo Horizonte.

Henrique picked me up in his car and took me out to the Mendes corporate complex, a handsome, sprawling estate with low white buildings hugging the green tropical landscape. The seasons are reversed there so though it was summer in NY, it was winter in Brazil.

I received a more thorough briefing on the company and the specifics of our research tour. Mendes had some eleven subsidiaries spread across the country and serious changes were underway in Brazil. As it was explained to me, the government owned the steel industry; however, it looked like it might go private and Mendes wanted to be in prime position to expand their existing steel operations and be a prominent player.

We toured several MJ plants that first afternoon so I could interview the executives. Everyone in management spoke English, obviously a prerequisite. Yet, everyone else spoke Portuguese, the official language of Brazil. It may be one of the more difficult languages on the planet. But, the people were terrific and so was the food on the tour.

After a second day of successful touring, I flew back to New York and wrote up my notes and various impressions so we could begin the design process.

The new Mendes Júnior Mark for this huge construction and steel manufacturing enterprise. 1994

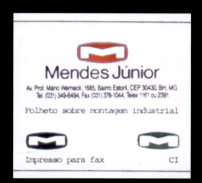

before: The old, existing Mendes Júnior Mark.

Gary Skeggs was to be the lead designer on the project, working with other staffers. After several weeks of exploration, we had developed a number of handsome symbols and logotypes. Typically, our firm did not offer the client many choices, sometimes only one. Perhaps, because of the unusual cultural concerns, we decided to present several different solutions, but with one strong favorite. This client felt a lot like family, and so I thought we could control the process and the outcome.

Mr. Mendes flew to New York for the preliminary presentation, and he came straight to our office and we immediately went into the presentation. He liked all the options we were showing. But, he was quite concerned about our favorite – it was a total departure from their existing mark; he felt it might be a hard sell to his people. Indeed, it was a radical change, but we felt that this change was required (their old logo looked flattened, like a heavy load of steel had been dropped on it).

We finally agreed we would move forward with the first choice. Over the next weeks, we applied the mark to a great variety of mediums, so that it became a tangible, fully understandable program. Then, I made arrangements for the presentation in Belo Horizonte and the entire management staff.

My trip down was a grind again, but I arrived in the late afternoon and was able to preview the full presentation that day. The staff session would be the following day when, hopefully, I would feel more human.

Some of the applications designed for our presentation of a complete and coordinated program for Mendes Júnior Group.

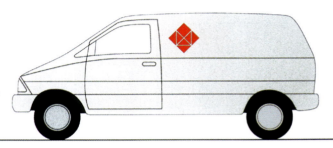

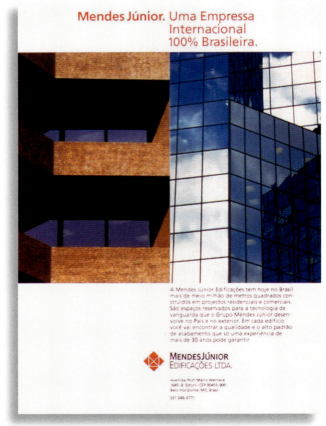

The audience was large and completely filled the auditorium space. Some of the Managing Directors had brought staff who would require translation. So the program was handled something like the UN, with headphones and several translators, to convert my English into Portuguese and French.

Mr. Mendes had requested that I present our credentials first, to show other programs we had done to add luster and weight to our Mendes redesign. In particular, he wanted me to show the full NASA program as it was well known and admired in Brazil, and he asked me to drop heavy names like DuPont, Standard Oil, and AT&T. That was the easy part and it was very well received.

Then I moved into the Mendes Júnior program, featuring the new Mark and the separate logotypes for the eleven divisions. At the first showing of the Mark, the whole room was set in motion. It was a very immediate and visceral reaction, not necessarily good. In retrospect, I believe the staff actually expected us to keep their old logo and just dress up the attendant graphics. That logo was dated and undistinguished, but it was theirs, and it had been for decades. There is nothing really unique here, I encountered this often when presenting a new program. There are always people who simply don't like, or understand change.

Of course, the mood improved as I trotted out all the applications, they began to appreciate that this was a broad, sweeping program, all Mr. Mendes had hoped for from the outset.

The verbal exchanges with the various managers were among my more interesting moments as a professional. The translations, the banter, it was all very lively. And after an hour or so, they were smiling, even laughing, with me. I wasn't the enemy now, and perhaps we could be an asset in the increasingly competitive marketplace that Mendes Júnior was facing.

I flew back to New York with approval of the new program. Over the next months, we would generate a handsome, comprehensive *Corporate Identity Manual* for Mendes Júnior Group. One was typeset in English, another in Portuguese. But when finished and shipped, the manuals were held up in Brazilian customs for over two months! The fact that they weren't created and produced in Brazil was a political issue, yet they finally cleared.

I enjoyed everything about this assignment and learned volumes about an extraordinary country. And at the end of a most rewarding project, I received this kind letter:

The highly professional performance, the dedication and full understanding of our business philosophy by you and your team have translated into the new corporate signature for the Mendes Júnior Group. This contemporary, strong symbol and logotype have been the reason for much admiration here.

The Corporate Identity Manual, with it's clear and objective manner, will be an important instrument for implementing and keeping our new high quality Corporate Identity.

It all goes to prove what we felt from the beginning, we picked the right firm. It was a pleasure to work with you and we hope to continue in the future.

With best regards,

J. Murillo Valle Mendes
President

The completed manual was put to immediate use once in the client's hands.

Left: a screen capture of a Mendes Júnior presentation – still honoring the mark and program today.

Good visual design can be powerful on its own, but it can also be exciting and memorable when animated.

Motion Graphics

20 second opening titles for NASA Public Service broadcasting, created and produced in 1977.

National Aeronautics and Space Administration   A great design program has many applications in various mediums. After the basic principals are established in the form of a style guide or manual, designers apply graphics to everything like web uses, packaging, vehicles, signing, advertising, television, and other appropriate media.

In the case of NASA, they required opening and closing titles for their educational science shows created for the public good.

I designed these titles as a NASA "special project," a couple years after our design program was introduced to the public. These were among the very first computer animations produced in New York City. The work was done at Dolphin Productions, and they were a real breakthrough at the time, done for a fraction of the cost of conventional animation!

The sound tracks were really special. The opening titles started with all synthetic music sound as the infinity symbol morphs into the planet. As the daVinci Man emerges, an acoustic cello overlays the basic track, creating a wonderful interplay – the past and the future.

For the closing title, we employed a hushed synthetic sound to enhance the space roll and soft landing of the logo.

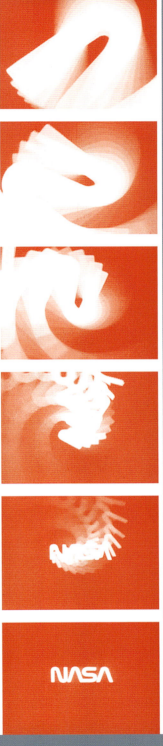

10 second Closing titles.

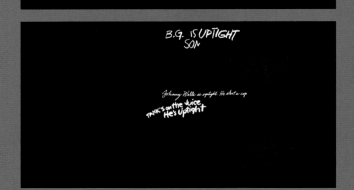

Invisible hands scrawl graffiti on a black screen until it is filled with violent messages from the film. 1968

Paramount Pictures The design firm of Gips & Danne was only around for five years, but they were a very productive five years. My partner, Phil Gips, and I worked on a great variety of design assignments, including many movie promotions over our last couple of years.

Our first work in this fascinating medium of movie promotions was on Robert Redford's first movie, *Downhill Racer*. It was followed by other films like *The Molly McGuires* with Sean Connery; The *Sterile Cuckoo* with Liza Minnelli; and the landmark promotion for *Rosemary's Baby* starring Mia Farrow. Most of these assignments were for Paramount and in collaboration with Steve Frankfurt, then president of Y&R Advertising.

Uptight is an American drama film directed by Jules Dassin, and was an updated version of John Ford's 1935 film *The Informer*. It is about a black man who shot a cop and is hunted down in New York City. Because of the ghetto quality of the film, I used graffiti as the messaging technique. Invisible hands scrawl harsh, animated messages on a jet black screen.

The sound track is very stringent and continues to build as the screen fills up. With the final graffiti, the soundtrack suddenly stops, leaving only the movie's logo in stunning silence.

30 second opening titles for "Science Report" a broadcast series created and produced in 1972.

The U.S. Information Agency This unique organization was responsible for delivering the American message to most parts of the world. USIA, which existed from 1953 to 1999, was dedicated to spreading "public diplomacy" and to extending America's influence to a multitude of countries.

Our assignment was to create opening titles for a broadcast series named "Science Report." And, because of the diversified audiences, the program's title appeared in a variety of languages.

The visual theme is "Genesis!" Each new subject was introduced with a light burst giving birth to the next image. These images included biology, chemistry, physics, ocean, and astronomy, etc. After the burst, images enlarge and exit the frame making room for the next subject. The sequence ends with the viewer being transported into deep space, inside a pattern of moving stars.

Former USIA Director Alvin Snyder recalled, "The U.S. government ran a full-service public relations organization, the largest in the world."

(Source: Wikipedia)

New York Jets for ABC In this ever-so-quick promo, typography is the name of the game. A long-standing tradition among Jets fans is to spell out the name of the team with great gusto. This sound track utilizes the entire stadium crowd as it roars out "J - E - T - S" as loud as possible.

Each letter spins out of the frame in rapid fashion until JETS appears with the football, which is quickly punted out of the frame as the ABC logo pops on.

It is really interesting to apply many of the same principles of graphic design to this realm of motion. Everything comes alive and offers new and exciting uses of our craft.

10 second promotion for the Jets football broadcasts. 1968

Featuring an array of diverse programs for huge organizations that prove that you don't have to be big, to perform big!

Super Scale

This section focuses on important assignments for oversized organizations. One might think that to produce projects and programs of a large scale, you would need a large staff. Not true; all of the projects shown here were generated by a firm ranging from 10 to 15 people. What is the difference, then, and how is this accomplished?

On the simplest level, it's the talent that really matters. We have always sought out talented designers and support personnel who are versatile, and who have good or even great range. These include inquisitive people who are well read and whose interests go well beyond the realm of specialized graphic design.

Additionally, I've believed in maintaining a stimulating mix of clients with very different missions and goals. This means both corporate as well as not-for-profit organizations. It also means an interesting roster of small organizations to counter the large ones we partnered with.

I'm often asked, "How can you handle these larger assignments?" Admittedly, it's difficult at first, but it gets easier with each outing. Through most of my career, I've tried to work with the CEOs of these organizations, or very close to the top, and this helps immensely. The higher up you go, generally, the smoother the process.

FAA

This redesign was started immediately after Danne & Blackburn completed the massive Department of Transportation program (including the "Trans" custom font). Langhorne Bond was FAA Administrator and the entire effort was accomplished in less than a year, considered a fast track for such an ambitious program.

*top: Exterior and interior signing
for FAA Technical Center, New Jersey.
above: Part of their diversified fleet
of two dozen different types of aircraft.*

left: Graphic Standards Manual. 1980

*Paint schemes were developed for each
type of aircraft in the fleet. Detailed
guidance – upper right – included top
and bottom of each type. 1980*

Advanced Materials Conference 24-25 October 1988

DuPont, Celebrating the past, Shaping the Future

The 50th Anniversary of Nylon and Teflon

*Many special events were staged during
the Anniversary year, including this
international Science Symposium.
Here, the poster / mailer for the event.*

Du Pont
Shaping the Future

DuPont invited our firm to create
a design / communication program
to celebrate the 50th anniversary
of Nylon. This orchestrated
program previewed in early 1988,
and ran for the entire year.

*Our Annual Report design also carried
the 50th anniversary theme, with a
milestone timeline running throughout
the editorial section.*

AT&T Bell Laboratories wanted some plaques designed to honor their science and research personnel. After considering the project for some days, I convinced the client to do something more memorable and aesthetic. These personally inscribed sculptural objects are the result, each given with significant monetary awards.

above left: The Distinguished Technical Staff Award is for sustained long-term achievement.
right: The Fellow award is presented for a one-time research discovery. 1982

AT&T

AT&T Consumer Products
1983 Distinguished Technical Staff Award
Donald S. Kennedy

*After the success of the Bell Laboratories
Awards program, AT&T asked us to
design a special piece for their Consumer
Products division. This award was
equally well-received. 1983*

For many years, our firm also created Annual Reports for this enormous corporation. With a print run of some 1.5 million copies per edition, it was considered the largest Annual of its time.

A spirited Annual profiling the talents of their many valuable employees.

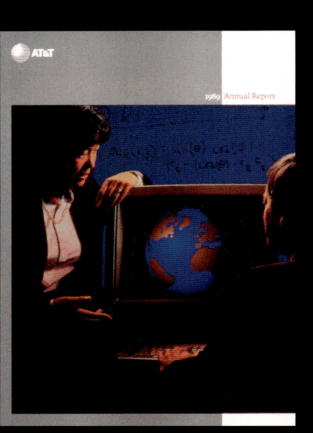

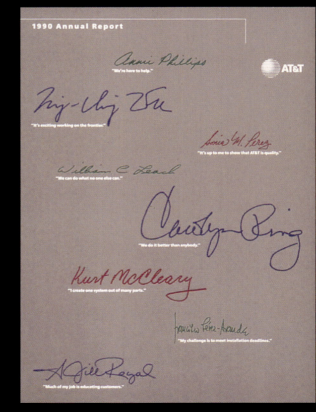

Our first Annual for AT&T featured their considerable global operations and plans for the future.

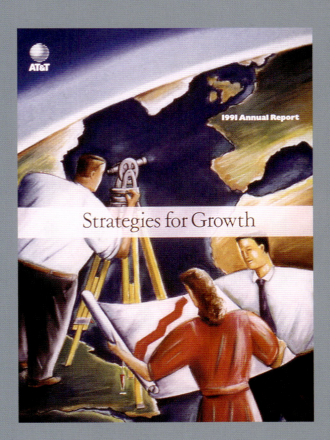

As the title implies, this Annual focused on the expansive growth that AT&T was experiencing at the time.

The Company had invested heavily in networking and they were able to showcase advancements in technology.

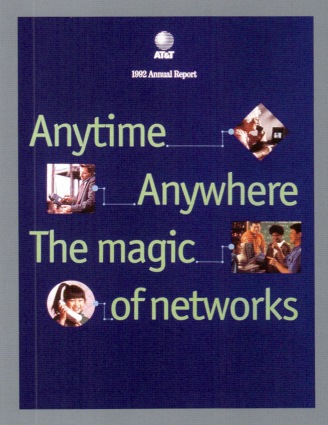

Gips & Danne designed many movie
promotions, most for Paramount Pictures,
and in conjunction with Steve Frankfurt.
Here is the landmark promotion for the film
Rosemary's Baby starring Mia Farrow
– for those unfamiliar she gives birth to the
devil. On the right is the theme poster
for the movie, now an iconic image. 1968

Paramount

Pray

*Prior to the film's opening, teaser
bus posters appeared. The first one (above)
and the second one, two weeks later.*

Pray for Rosemary's Baby

This handsome garden environment is lined with numerous retail shops. We designed the mark and packaging for Atrium on Public Square.

Standard Oil was a client for many years, so when they built their new headquarters in downtown Cleveland, we were pleased to help introduce this major building to the community. Tours were scheduled over a two-week period to accommodate the considerable interest of the citizenry.

Standard Oil

above: Media Kit and Tour Program for Opening Week celebrations. 1986

right: A traveling exhibition portraying the entire history of Standard Oil. It was installed for Opening Week and then traveled to schools and public spaces.

Our relationship began in 1983, and it survived three name changes – first as Sohio, then as Standard Oil, and finally as BP America. Below are several of our Annual Reports for the Company.

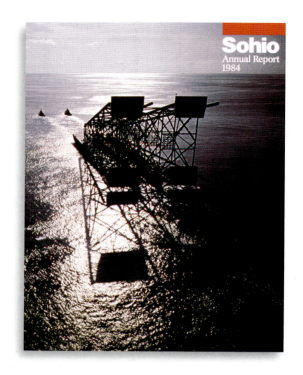

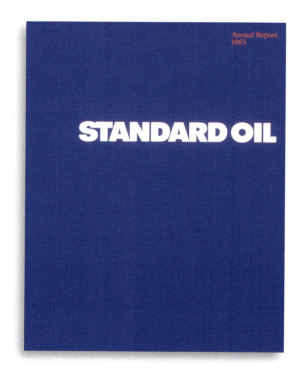

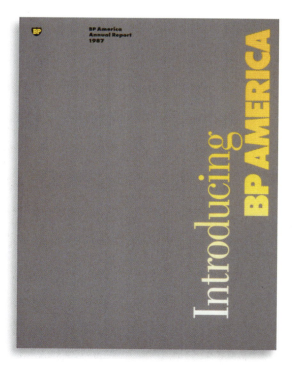

No matter what you label it, one of the
most interesting areas of design is branding.
It allows you to create an important
and long-lasting foundation for the client.

Marksmanship

When I entered the field decades ago, a company's graphic name was its "signature." It was a nice term which suggested something personal and unique, just what any organization would want to achieve.

Later, sometime in the mid-fifties, the term "identity" came into widespread use. This was a very descriptive way of classifying the graphic presentation of a company, and it also was appropriate and effective.

In more recent years, "branding" has become the universal term for an organization's visual presentation to the public. This combined with the term "strategy" have become so pervasive that it is almost cliche. But, giving credit where it is due, it is effective if not overused.

In my own practice, I now use the term "brand identity," as an amalgam which is both descriptive and complete.

There are many programs in this book which serve as examples of total brand identity, starting with NASA. The following pages feature other programs which have a strong and convincing mark at their foundation. Here are a variety of organizations both large and small, corporate and not-for-profit. They illustrate the diversity of our practice and our wide-ranging interest in communication for the public good.

*Total program for the largest
public utility in the U.S.
with various generation methods
including hydro power.
New York, New York. 1983*

*Anniversary symbol
for America's oldest music school
settlement.
New York City. 1994*

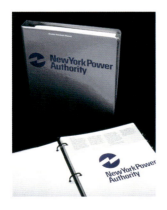

*Mark to introduce a new
motion picture producer
and distributer.
Dallas, Texas. 1961*

*Comprehensive program for a
developer / manufacturer
of surgical instruments for the eye.
Armonk, New York. 1987*

Paula Schwed

ver·ba·tim

*Editorial services
for all media, with the phrase,
"When Words Matter"
Decatur, Georgia. 2004*

Kid
Knowledge

Kid
Knowledge

Kid
Knowledge

*Marks for a breakthrough
teaching method of
K-2 science instruction.
Armonk, New York. 2009*

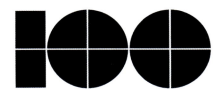

*100th anniversary symbol for
the oldest chapter of
Printing Industries of America.
New York City. 1964*

*Three distinct Swiss restaurants
under one roof
at Rockefeller Center.
New York City. 1970*

*Elegant photography books
featuring vineyards and wine making.
Napa, California. 2012*

*Innovative school for
children with behavioral and
learning deficiencies.
Brewster, Massachusetts. 2006*

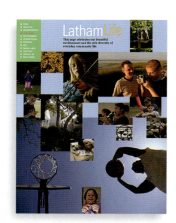

*Software to connect
government procurement with
outside vendors.
Austin, Texas. 1996*

*YPO is a worldwide coalition
of dynamic young chief executives.
New York City. 1984*

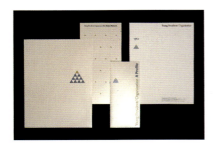

TheBlueshirtGroup

**Springboard
Biodiesel**

*Preeminent firm specializing
in Investor and Media Relations.
San Francisco, California. 2004*

*Progressive use of waste bio
material to create alternative energy.
Chico, California. 2008*

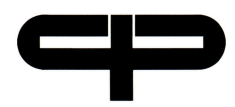

*Firm of accomplished
architectural designers.
Dallas, Texas. 1959*

*For learning-challenged students,
with our phrase,
"Caring. Teaching. Guiding."
East Sandwich, Massachusetts. 1999*

25th anniversary mark for the
National Aeronautics
and Space Administration.
Washington, DC. 1984

Capital Campaign with the phrase,
"Spreading Knowledge,
Strengthening Community"
Osterville, Massachusetts. 2004

*Founded in 1838, Atlantic is a
leading provider of
Marine insurance to the world.
New York City. 1987*

*Theme of the Annual Gala
for the school.
East Sandwich, Massachusetts. 2003*

*Some of the rotating marks
for an interactive data center serving
all constituents of the Great Lakes.
Ann Arbor, Michigan. 2022*

Posters are among the most powerful mediums, ever. Yet, they were never fully utilized in the United States.

Poster Power

European communications have been poster-driven for many years, and there is a lengthy legacy of outstanding poster design. The best designers have produced memorable images and messaging which were often the main mode of advertising. But even that has changed as the internet has become more dominant. The medium never gained real popularity in America, but here are a select number of examples from my own career. (Other posters appear throughout this book.)

From a series of six-foot posters designed to anchor a one person exhibition at The Museum of Art, Oklahoma University. 1966

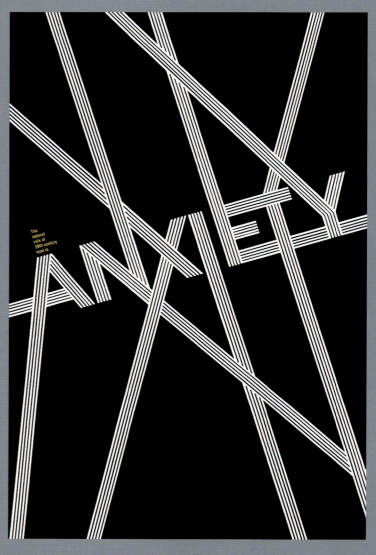

Norman Mailer:
"Anxiety is the natural role of 20th Century man."

Adolf Hitler:
"The one means that wins the
easiest victory: terror and force."

Honore de Balsac:
"Bureaucracy is a giant mechanism
operated by pygmies."

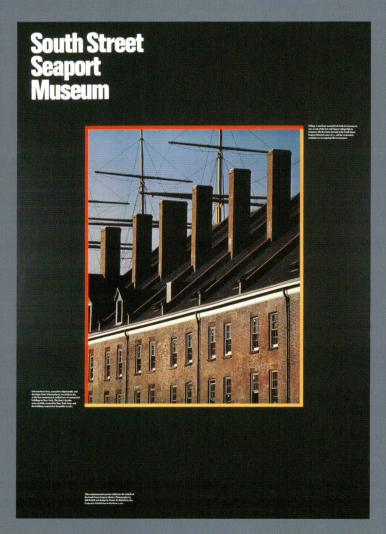

*Commemorative poster for the grand opening of
the South Street Seaport Museum on the
East River in historic Lower Manhattan. 1983*

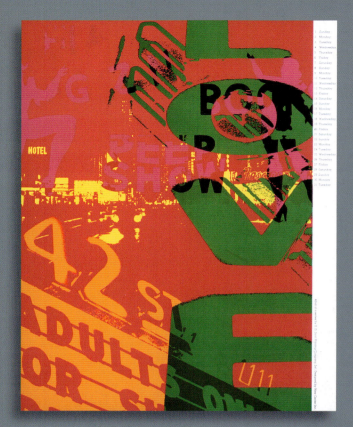

*42nd in series of poster calendars for
SD Scott Printing Company in NYC. 1972*

MAGNUM

NEW YORK PARIS

Founders
Robert Capa
Henri Cartier-Bresson
George Rodger
David Seymour

Members
Eve Arnold
Bruno Barbey
Ian Berry
Werner Bischof
René Burri

Cornell Capa
Bruce Davidson
Elliott Erwitt
Leonard Freed
Paul Fusco
Burt Glinn

Burk Uzzle
Philip Jones Griffiths
Charles Harbutt
Erich Hartmann
David Hurn
Richard Kalvar

Josef Koudelka
Guy Le Querrec
Erich Lessing
Mary Ellen Mark
Costa Manos
Wayne Miller

Inge Morath
Gilles Peress
Marc Riboud
Marilyn Silverstone
Dennis Stock
Burk Uzzle

15 West 46th Street
New York 10036
Telephone: 541-7570
Cable: MAGNUMFOTO

2 Rue Christine
Paris 75006
Telephone: 325-90.09
Cable: FOTOMAGNUM

Dear Senator,

The Senate may vote soon on S.2387, the bill to break up the oil companies. Only an aroused electorate, barraging Congress with mail, can assure its defeat—and help assure America's energy future.

So write today, while there's still time. For 13 cents, you can help stamp out political opportunism and protect jobs.

Divestiture is your problem. Let your Senators and Representative know how you feel. Write. Right now.

Mobil

To encourage Mobil employees to write to their representatives regarding legislation directly affecting the company, 1975

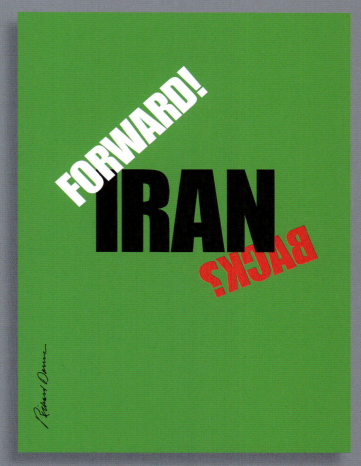

Major invitational exhibition staged
in New York City prior to special
elections being held in Iran. 2009

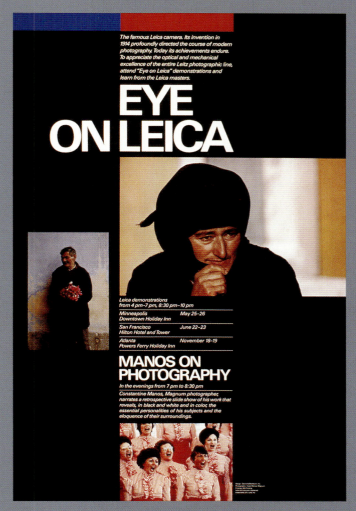

For Leica Camera AG, an exhibit
featuring the stellar photography of
Costa Manos. 1975

Richard Darroe

Simpson

Be sure you
have enough
fuel onboard
for your flight,
plus... **15** minutes
reserve fuel
for take-off,
plus... **45** minutes
reserve
fuel at
destination

U S Department of
Transportation
**Federal Aviation
Administration**

*Aviator Safety Regulations for the
Federal Aviation Administration, as part of
a total redesign of the Agency. 1981*

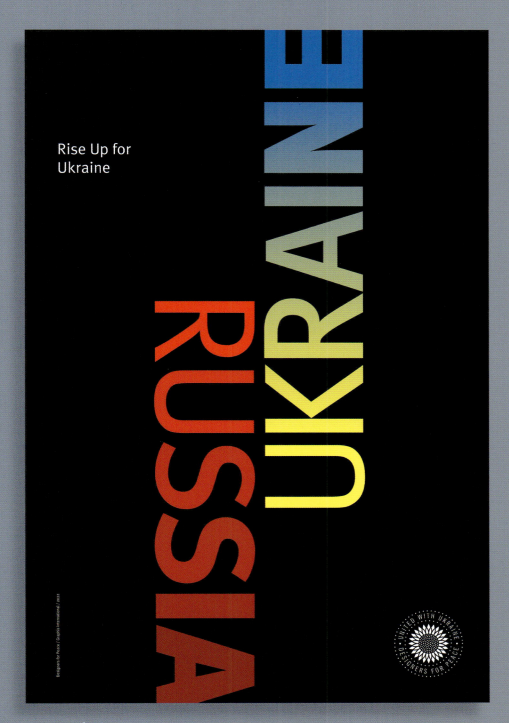

Rise Up for
Ukraine

Designers for Peace.
My design for this important campaign
conducted by Graphis Magazine. 2022

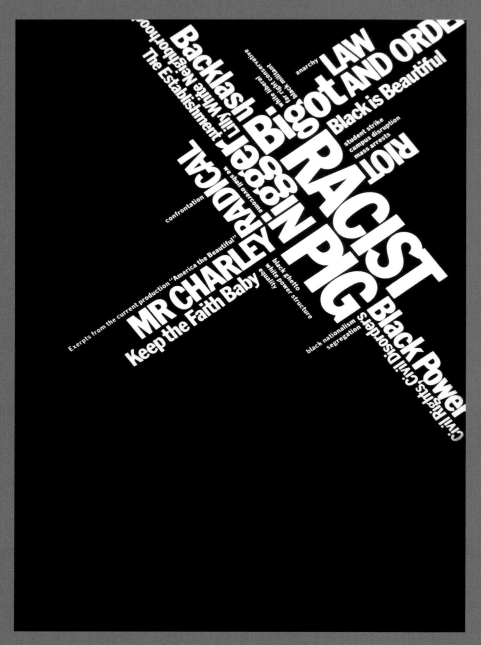

*For "Black and White," an invitational
exhibition by Print Magazine.
I used quotes excepted from the play
"America the Beautiful." 1970*

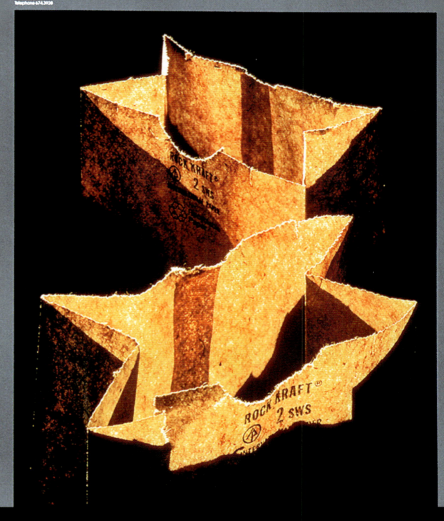

A poster / mailer for Jim Barber,
an accomplished studio photographer
in New York City. 1982

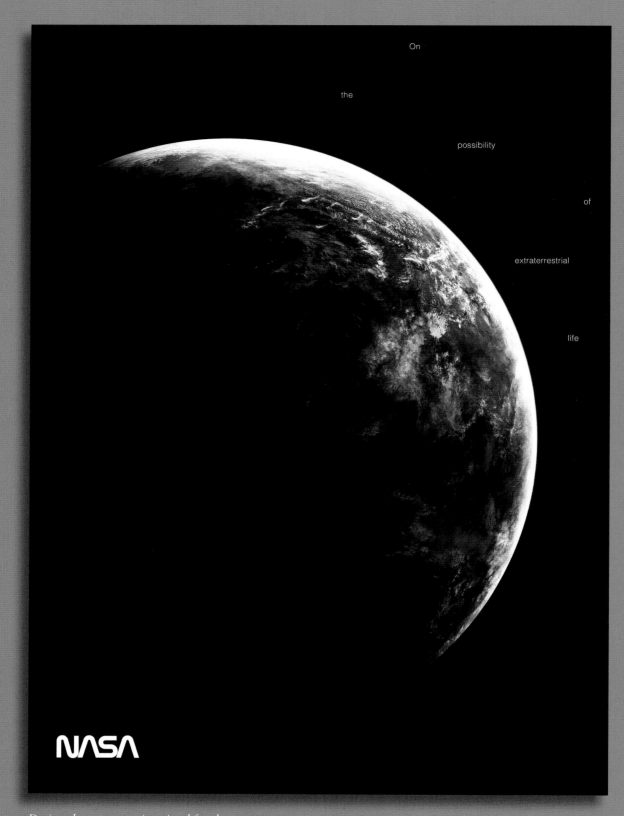

On

the

possibility

of

extraterrestrial

life

Designed as a supporting visual for the
1974 NASA redesign presentation. Then reprinted
in 2023 by HOLOCENE magazine.

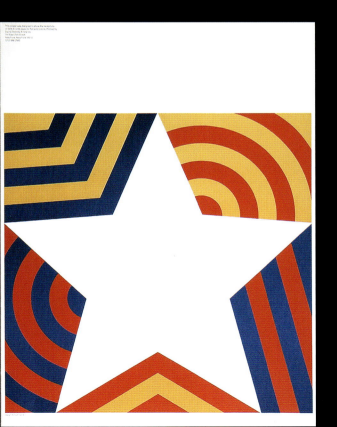

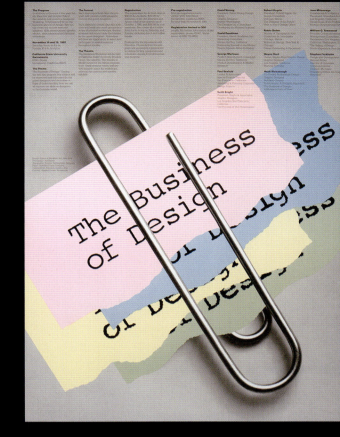

*The Art Directors Club of Sacramento
announcing their inaugural
Business of Design conference. 1981*

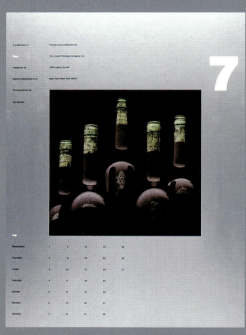

"A Collection of Time"
With platinum silver backgrounds, this
series explores unusual aspects of
time. Created for SD Scott, a premier
printer in New York City. 1984

Having a solo exhibition is among the most satisfying experience one could ever experience. It's a ton of work and well worth it.

Solo Shows

Here is a sampling from the Fall 1988 exhibition at the
Museum of Art, University of Oklahoma. The entire
show was shipped and presented again at my alma mater,
Oklahoma State University, and Gardiner Art Gallery,
in the winter of 1989, with a Master's class for students.

Magnum Photographers:
Anniversary poster.

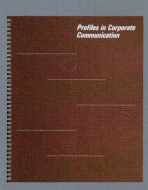

Simpson Paper:
Corporate Communication.

Bell Laboratories:
Scientific Achievement Awards.

Potlatch Corporation:
Annual Report.

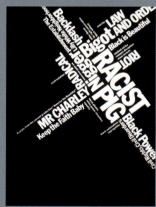

Print Magazine:
"Black & White"
invitational exhibition.

Harvard Business School::
75th Anniversary publication.

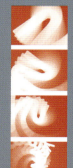

NASA:
Closing TV titles.

SD Scott:
"A Collection of Time" calendar.

NYPA:
Design Standards Manual.

Seagram Company:
Fine Wine Annual Report.

Mead Paper:
"New York New York."

Westinghouse Broadcasting:
Album of World War II victories.

NASA:
Graphics Standards Manual.

Federal Aviation Administration:
Total redesign of the Agency.

   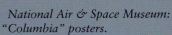   

National Air & Space Museum:
"Columbia" posters.

more >

Swiss Center Restaurants:
Architectural graphics.

State University of New York:
Quarterly publication.

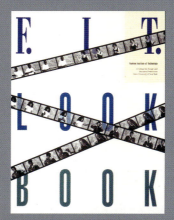

Fashion Institute of Technology:
Recruiting book.

Seagram Company:
Portfolio of Premium Brands.

Westinghouse Broadcasting:
"The Music Goes Round & Round."

Atlantic Mutual:
Design Standards Manual.

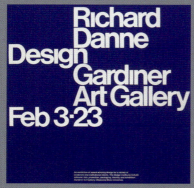

Oklahoma State University:
Campus poster.

*The world has gotten smaller and designers
are now working in cross-cultural environments.
Here's an unusual example of what can go
right – and wrong – in the process.*

Arabian Episode

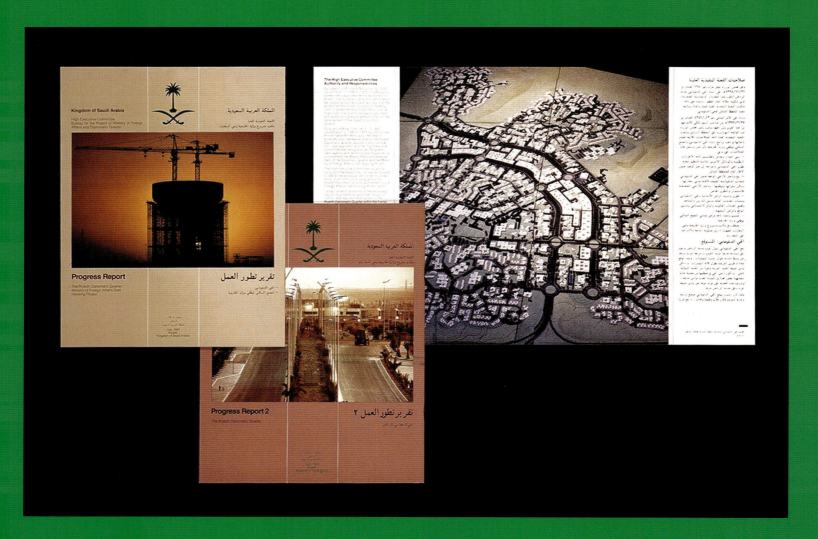

We designed a series of "Progress Reports" to track the development and construction of the massive Royal Diplomatic Quarter complex (small section is shown in the model above). The overall project was a huge undertaking – a new city for 50,000 international diplomats – which required a multitude of planners, architectural firms, and landscape specialists.

April, 1982

The trip over was a struggle, 18 hours door-to-door. I flew to Rome first and then took a Saudi Airlines flight to Riyadh, arriving in the late afternoon. I was dead tired, and loaded with apprehension from stories about baggage checks and detentions for the least little thing. Not too likely in my case, as I was traveling with a letter from the Foreign Ministry which explained my presence in Saudi Arabia. I finally cleared inspection but felt completely alone. Men and boys swarmed around me, all wanting to separate me from my luggage and take me to my destination.

Emir Habibi was my contact, a great guy who I had met in New York, but there had been some miscommunication. I managed to reach Emir by phone and we agreed to rendezvous at his brother's apartment, who was out of the country at the time. The two brothers were American citizens, lived on separate East and West coasts, and both were hired as consultants for this massive Diplomatic Quarter project.

The apartment was located in a two-story building and was quite attractive and fairly comfortable. It had a refrigerator loaded with a kind of pita bread and a large amount of cooked chicken, though I wasn't sure it was meant for me. It was a fairly simple place with one true asset, a serious and sizable collection of LPs, including tons of Mozart, Bach, and other classical staples.

Emir got me settled, and he notified his people that I had arrived safely in Riyadh. Then, he arranged to pick me up the next day to start the protocol, the patient waiting before being called for my presentation.

On the second day, I was ushered into the elegant offices of the director general of the Diplomatic Quarter project. There were several other men in the room all dressed in handsome robes appropriate for their kingdom rank. Not the usual guys in ties. I was very respectful of them and the situation, but I had been hired as a expert consultant and I needed to be confident and be myself. They reviewed my design presentation of the first "Progress Report," asked the right questions in perfect English, and responded positively. They were delighted.

Most traditional Arabic/English publications read back-to-front with the Arabic first – the English translation starts at the other end and the two meet in the middle. It's a very awkward format. My book had the Arabic and its English translation facing each other, in a mirror-image mode. The client thought we had "reinvented the wheel" and were generous in their praise. I exited the building with Emir, both of us pleased and relieved. If you are going to travel half way around the world to present, much better to get a win!

A spread from "Progress Report Number 1" with mirror image text.

The Diplomatic Quarter itself was breathtaking and unprecedented in both style and scope. Each country had selected one of their leading architects to design its embassy. The comprehensive Master Plan was done by Germans – figures – and Landscaping by English – also figures. The infrastructure was being built by Korean workers who, I was told, were obligated to fight and defend should Saudi Arabia be invaded during this time. The schools, public spaces, athletic clubs, etc. were all designed by noteworthy firms from around the world – Unbelievable!

So you might be wondering: why was the Saudi Ministry publishing these books to begin with? The Old Diplomatic Capitol was located in Jeddah, on the Red Sea coast. Saudi Arabia is a strict country, though Jeddah had a looser social and religious code (one could actually get a cocktail there). But Riyadh is the religious capitol of this Islamic Kingdom, by contrast, Jeddah was Las Vegas! The world's diplomats really didn't want to move to Riyadh, presenting an enormous public relations problem for the Saudis. So, these publications were conceived to paint the new Quarter in the best possible light.

Enormous water towers under construction.

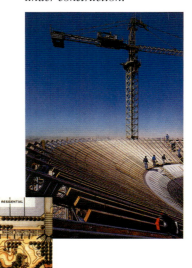

That evening, word came over the radio of the mass shooting in a Palestinian Mosque, and all hell broke loose! Air space was immediately shut down and a full military scramble ensued. These events could go on for days – or weeks – I had come oh so close to getting out of town but now all was uncertain. I called my wife Barbara, in New York, to give her the bad news, and I proceeded to spend a very restless night, hardly sleeping.

Detail of the landscape plan for a neighborhood.

The next day, I checked the radio for any new information on the military crisis. You couldn't call the airlines, you had to go out to the terminal at night and stand in line with other desperate people, only to find out there were no commercial flights.

There were hundreds of consultants and workers employed on this huge project. Someone heard about my plight, and I was invited to an informal dinner party at the home of an Australian architect. With nothing on my social calendar, I said yes.

I was picked up at a certain hour and driven to a very large private residential area. All the consultants lived there and were allowed to behave like they were back home, while Saudi officials looked the other way. A consultant would "sign on" for a stint of specified length, usually 6 months. With nothing to do in the desert, 6 months would be an eternity. At the party, everyone was commiserating with me about being pinned down there. They could truly identify – though they made good money – they were also pinned down. There were professionals from around the world and everyone was letting go.

Model of the Diplomatic Club, a social center.

below: The fully finished Club.

You couldn't bring liquor into the country, so one of the guys had fashioned a crude wine. Starting with dark grape juice, he had added yeast to create a powerful home brew in a hurry. It was not pretty, but no one else minded because it yielded the desired effect – and no one died.

I've never seen so many unhappy people having such a wonderful time. After this spontaneous evening of fun and games, someone took me back to my lonely apartment.

After a couple of fruitless nights at the air terminal, I was directed by some friendly military personnel to the front of a line where arrangements were made for my departure the next day. I'll never know who pulled the strings… but I showed up way ahead of schedule still not believing it would happen. But, as I boarded the Saudi plane and took my seat, I began to think 'this might work!'

To my surprise, two landscape architects I had met at that party also boarded the plane; they were leaving their contracts early with no intention of coming back.

Our plane finally did take off, it was surely the nicest liftoff I ever experienced. We were actually heading towards Italy – the Promised Land!

We landed in Rome and took a cab to the Hotel Superiore, those consultants obviously hadn't made plans, so they decided to stay at the same hotel. After checking in, I went upstairs to unpack my camera and race around the "Eternal City." The other two guys went straight to the bar – they had escaped their Saudi contracts – and were proceeding to "make up for lost time."

I scurried around the city, shooting as fast as possible. The light was incredible and I thought:,"I've never seen a more beautiful place." As darkness approached, I somehow found my way back to the hotel. I walked through the lobby and past the tiny bar – Sure enough, my new pals were still celebrating their good fortune, still doing their part to drain the place.

I called Barbara, I couldn't wait to get home, and the next day, I was flying back to New York – grateful for many things:
- The Diplomatic Quarter project;
- my first taste of Italy;
- and the feeling I had truly dodged a bullet.

My shot of the famous "Spanish Steps" in Roma.

Another of The Vatican, in very late afternoon.

Current landscaping and buildings in the Royal Diplomatic Quarter.

My podcasts and lectures to Web designers confirm they have a great deal of respect for design rendered with ink on paper, and they often find it inspirational.

Paper + Print

There is nothing like the smell of ink in the press room, it is embedded in memory!

The following are projects designed for the finest paper companies and printers of the time.

Strong concepts and ideas can transcend any medium – including print.

Strathmore: "Strathmore on Opaques"
We helped develop breakthrough ink technology,
then printed opaque offset inks on the Grandee
line of papers to achieve silk screen effects. 1969

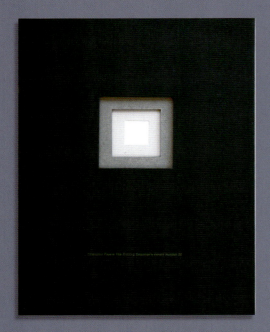

Champion International: "Herald"
A periodical, this was the "In"issue. One story,
"Inside Man Out," is about a Plant Manager

Beckett Paper Company: "Cheese Inc."
The client invited me to create a fictitious company
for their ivory colored stock. Cheese Inc. is
the result which won Letterhead of the Year. 1967

Mead Library of Ideas: "Printer's Choice"
An invitational exhibit for printers who selected
their favorite projects of the year. 1969

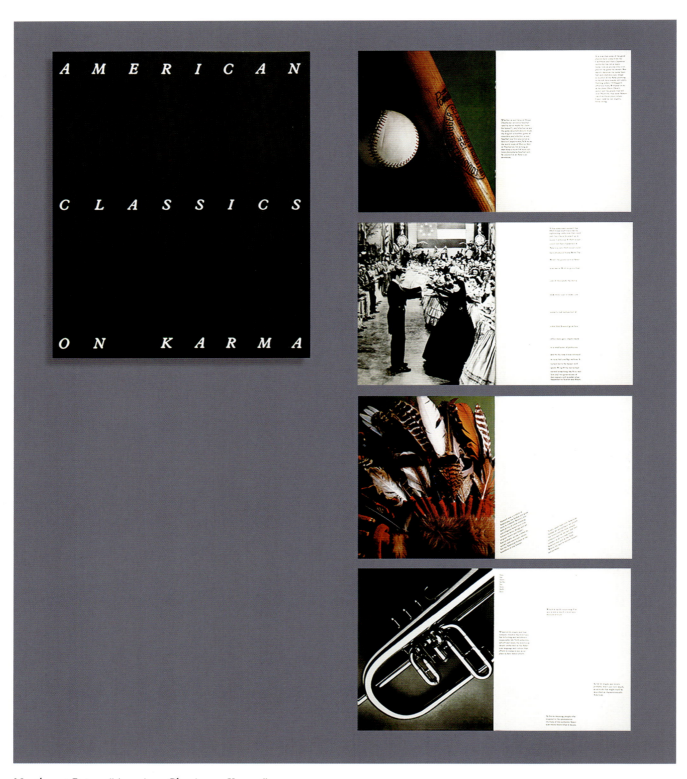

Northwest Paper: "American Classics on Karma"
Brochure to introduce Karma, a new matte sheet.
Featuring All-American subjects like – baseball,
movies, Indigenous arts, and jazz. 1980

Kimberly-Clark: "Winter Seen Through the Eyes of Children."
A tour de force workout of the entire line of K-C papers.
Seven different photographers were used, and children wrote
the captions to the photographs. 1970

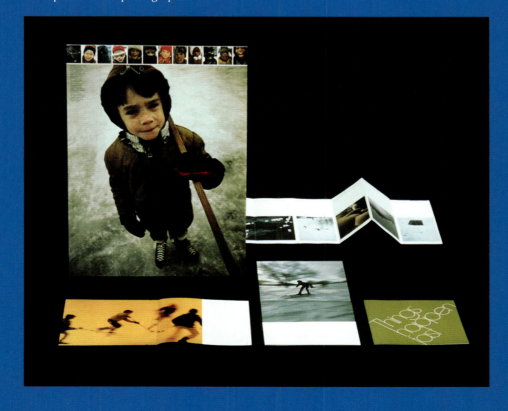

Mead Paper: "New York New York"
No words, just my graphic
impressions of the "Big City,"
on premium grade enamel. 1973

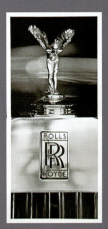

Simpson: "Profiles in Corporate Communication"
A promotion featuring a variety of Simpson papers
and profiling great programs inc: Polaroid, Apple, Espirit,
Lomas Nettleton, and United Technologies. 1985

SD Scott: "The Style Calendar"
From an ongoing series of wall calendars conceived
to associate the printer with sophistication and
high quality. Front: small perfume bottles. 1986

Rapoport Printing: "Hand Crafted"
For a printer renowned for craftsmanship,
some of a series featuring hand crafted subjects:
One-of-a-kind Rolls Royce hunting car;
The Stradivarius violin;
Unique canvas aircraft of WW I.
Dust covers enhance each subject. 1965-67

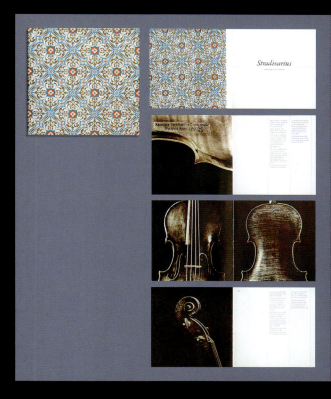

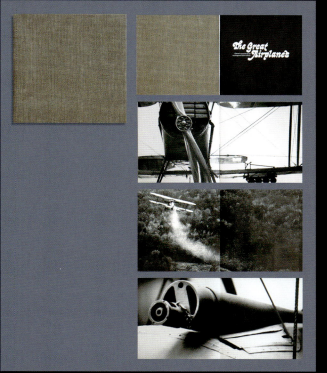

Wilmans: "The Art of Printing"
A boxed set of 12 individual mailers, each featuring
a milestone event in the history of printing.
For Dallas' best printer, I created the concept, did
the research, and carved the woodcuts. 1962-63

For decades, Annual Reports were an organization's primary marketing and financial reporting tool. They were a force and we were a prime player in this unique medium.

The Bottom Line

On the following pages are some of the award-winning Annual Reports we designed and produced over many years. Note that there are multiple years for each client, not one-off solutions. We are proud that we were able to have sustained relationships over time – this fact speaks to the ability of a firm to meet the client's goals while remaining friends in the process.

Potlatch

1979 Annual Report

Shown here is the 1979 edition of the successful series of Annuals for Potlatch, a forest products company.

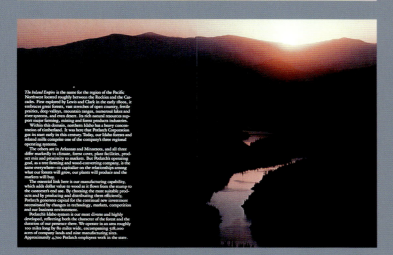

The interior pages detail the vast Potlatch operations in what is known as the "Inland Empire."

Other editions of the much honored Potlatch Annual Reports.

The Seagram Company Ltd. Report for the twelve months ended January 31, 1987

Seagram in the World of Fine Wines

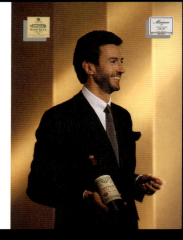

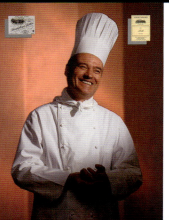

The Seagram Company used themes in all of their Annuals. Here featuring their role in "The World of Fine Wines," utilizing global personalities. 1987

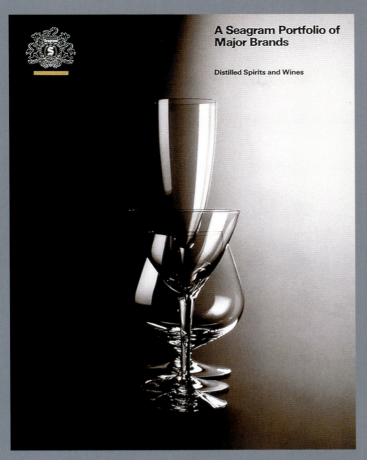

This Annual showcased Seagram's many famous brands which were accompanied by interesting background lore. 1988

Other editions featured their brown liquors and new acquisitions.

*A small portfolio of Annual Reports
are shown here; ARs for other clients
appear throughout this book.*

AmeriTrust Corporation

Bristol-Myers   *American Home Products*

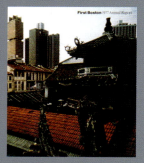 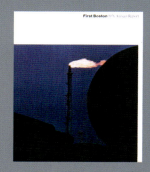 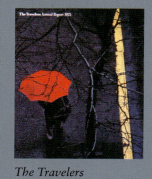 

First Boston   *The Travelers*

 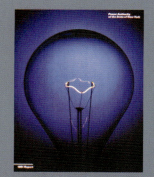  

New York Power Authority

We operate in a world of specialization and the design profession is no different. But our firm has always worked in many mediums, here is a sampling of design – outside of category.

Popourri

Owens-Illinois:
Banners in the corporate headquarters

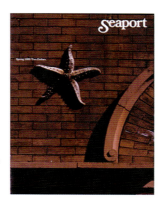
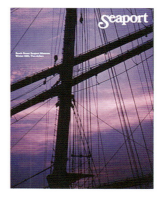

South Street Seaport Museum:
Seaport, *a magazine to build interest in the*
Museum, located in historic downtown
Manhattan, prior to its opening. 1979-81

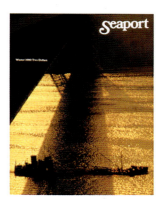

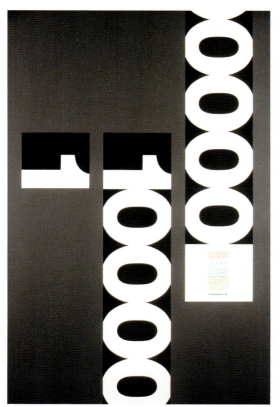

Time Life International:
Accordion piece to announce their
one-millionth subscriber, the
message repeated in five languages
on the final panel. 1968

Barbara & Richard Danne

Greeting card for professional and personal use. 1967

Pyramid Wrapps:
Mark, menus, and architectural
graphics for a prototype healthful
restaurant in New York City. 1997

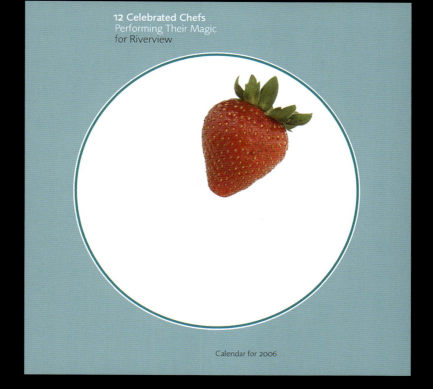

Calendar for 2006

designed for an annual
Chefs. This edition
their magic." 2006

Science of the **Brain**

GraviScience:
Provocative publication exploring
the brain research being conducted
by international scientists. 2005

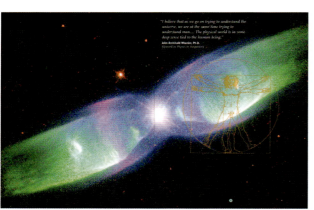

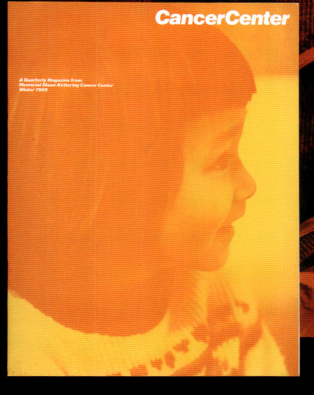

Memorial Sloan Kettering Cancer Center:
Quarterly magazine with the myriad
developments of this leading medical
organization. 1989-81

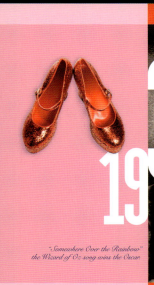
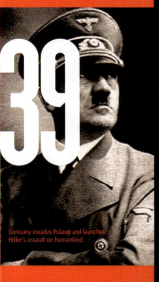

AIGA Centennial:
Invitational exhibition based on 100 years.
My assignment was the year 1939, when
"Over the Rainbow' won the Academy
Award, and Hitler invaded Poland. 2014

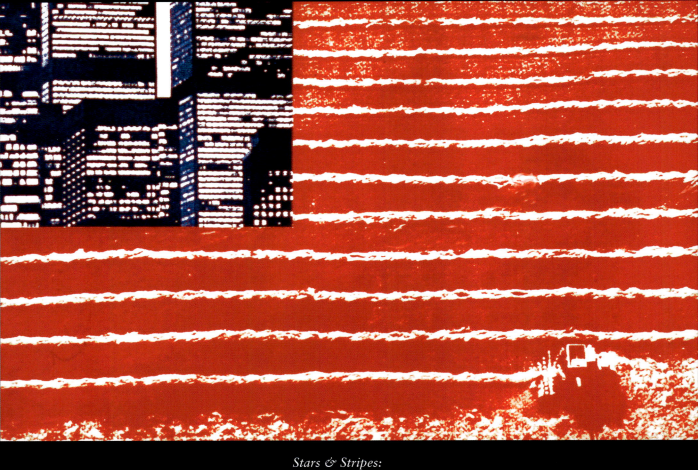

Stars & Stripes:
My (autobiographical) contribution
to an invitational exhibition
created by Kit Hinrichs. 1993

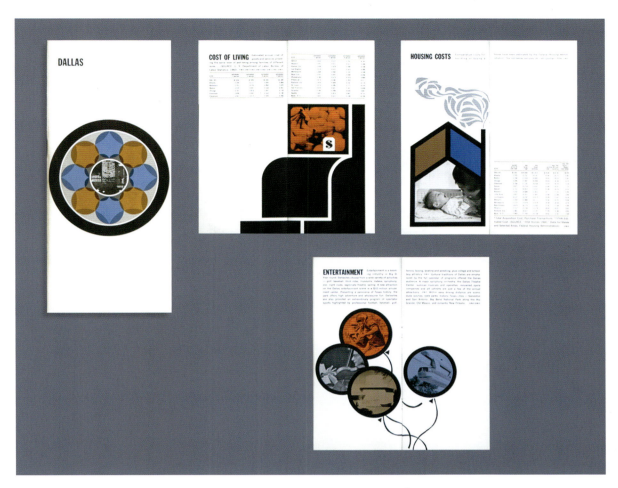

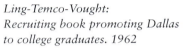

Ling-Temco-Vought:
Recruiting book promoting Dallas
to college graduates. 1962

Wellfleet Harbor Actors Theater:
Roll-out brochure spells
WHAT, the noted theater founded
by actor Julie Harris. 2006

Monday	Monday	Monday	Monday	Monday		3	10	17	24	31
Tuesday	Tuesday	Tuesday	Tuesday	Tuesday		4	11	18	25	
Wednesday	Wednesday	Wednesday	Wednesday	Wednesday		5	12	19	26	
Thursday	Thursday	Thursday	Thursday	Thursday		6	13	20	27	
Friday	Friday	Friday	Friday	Friday		7	14	21	28	
Saturday	Saturday	Saturday	Saturday	Saturday	1	8	15	22	29	
Sunday	Sunday	Sunday	Sunday	Sunday	2	9	16	23	30	

S. D. Scott:
A NYADC Gold Medal calendar
with a "Circles" theme,
for a premier NY lithographer. 1980

Atlantic Mutual:
Inaugural issue of a quarterly magazine
featuring new products from this
venerable insurance company. 1988

AIGA:
Logotype and materials for a star-studded
gala honoring medalists and other awardees.
top to bottom: Web, fold out Invitation,
and 72-page program. 2015

McCarran Air Terminal:
Signing for the massive expansion of the
Las Vegas terminal. Served as consultants to
TRA architectural / planning firm. 1981

Transamerica:
Introductory brochure
for their foreign exchange
service. 1971

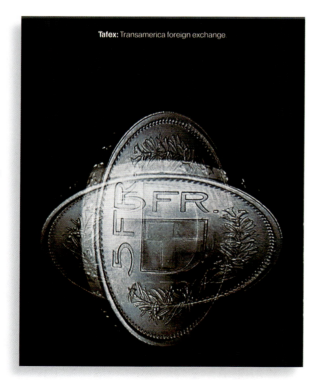

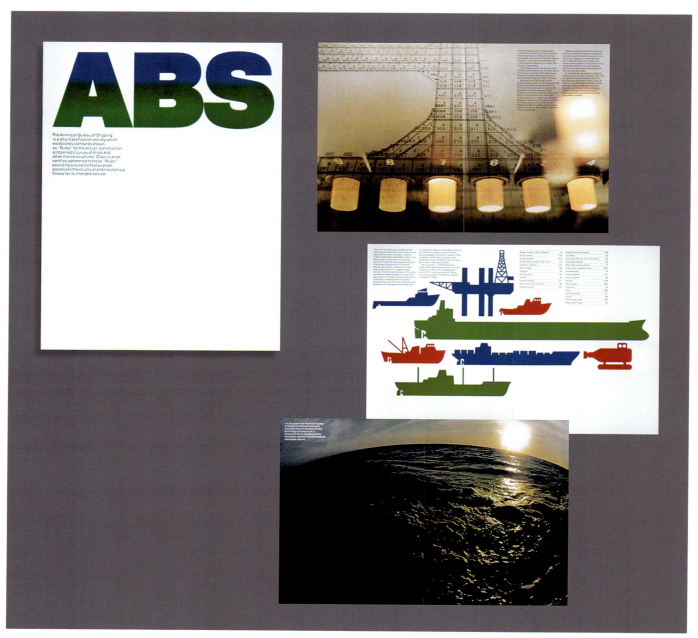

American Bureau of Shipping:
Capabilities brochure for an organization which designs and classifies
all types of naval vessels. 1972

The Drawing Board:
"Christmas in the City,"
one of many greeting cards for
this Dallas publisher. 1960

GRID 1234567891011121314@15

18 points   18 proofs   18 pixels   18 years   GRID@18

GRID@XX

Grid Typographic Services:
Text and design for a long-running
campaign for New York's best
typographer. 2005-16

a small school
built
on a **big** idea

the penikese island school

What Penikese is not

Where birds and boys find safe haven

The sophistication of simplicity

The Penikese idea

PENIKESE

Penikese Island School:
Roll-out brochure for a small
school for wayward boys
in trouble with the law. 2001

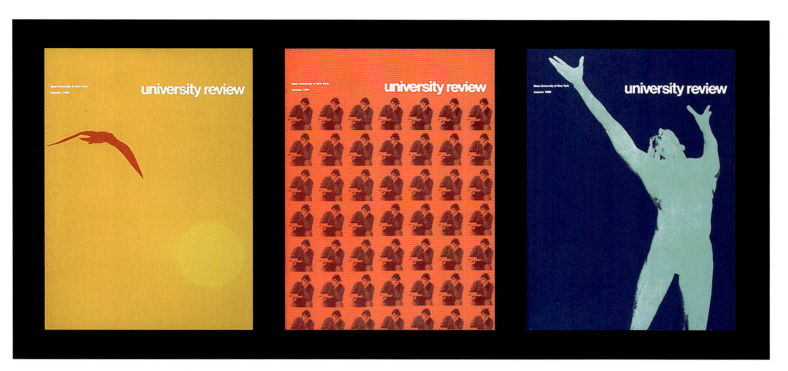

State University of New York:
A "think' publication for faculty & influencers.
left to right: Summer on the Adriatic,
Classroom boredom, Evolution. 1966-72

Boys Life Magazine:
A traveling photography exhibit
"Discovery – Images of Youth" previewed
in NYC before touring the US. 1972

Pratt & Whitney:
Publication Design Guidelines built
upon their circle heritage. 2001

Cyrano de Bergerac / EDMOND ROSTAND

Cyrano de Bergerac is Edmond Rostand's immortal play in which chivalry and wit, bravery and love are forever captured in the timeless spirit of romance. It is the moving and exciting drama of one of the finest swordsmen of France, a gallant soldier, brilliant wit, and tragic lover with the face of a clown. Captivating, bittersweet sadness and exquisitely polished language have made Cyrano de Bergerac the best-loved play in the literature of the stage.

THE BANTAM LIBRARY OF WORLD DRAMA

A new repertory of living theater chosen from the whole span of world drama. Authoritative editions of the classics... fresh modern translations ...important plays of our time... the latest experiments of the avant-garde... with introductions and commentaries by leading critics and scholars.

Cyrano
de Bergerac

Edmond Rostand
Translated by Brian Hooker
Complete and Unabridged

Bantam Library:
Covers for the paperback edition of
the literary classic Cyrano. *1965*

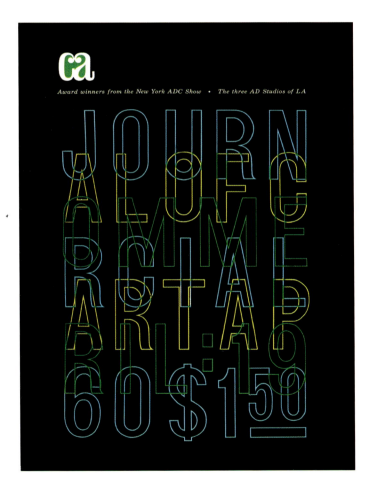

CA Magazine:
Early cover for this new publication,
while freelancing in Dallas, TX. 1960

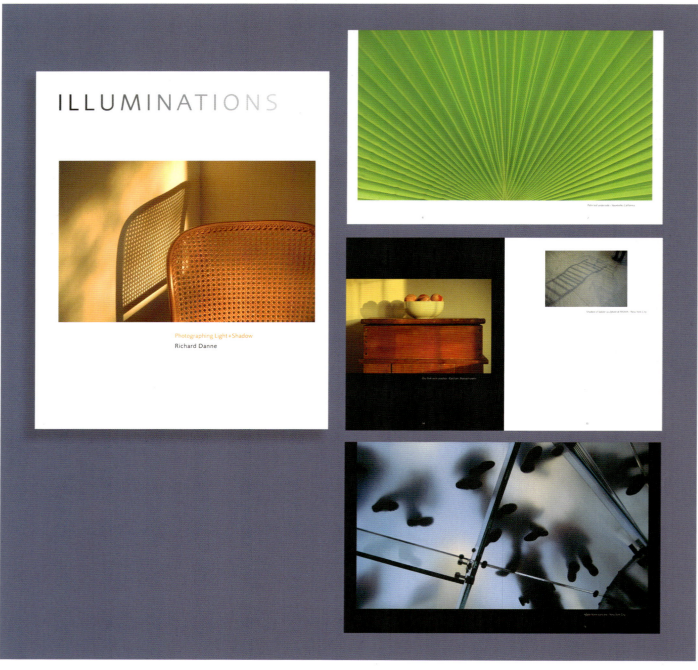

Illuminations:
Book of personal photography
from around the world, exploring the
effects of light + shadow. 2010

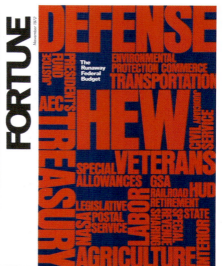

Fortune Magazine:
1972 Federal Budget with
each department
scaled to its allocation.

H&S
REPORTS

For the Firm of Haskins & Sells/Autumn 1968

Helping the arts escape red ink—page 26

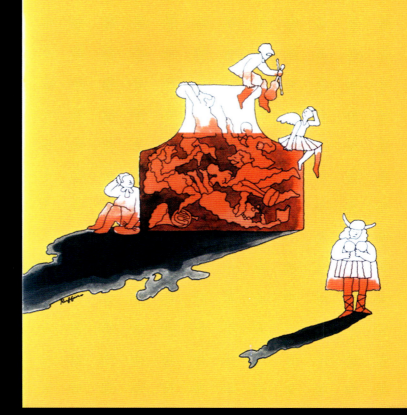

Haskins & Sells:
Consultant to this CPA firm for years.
Designed a 75-year history book, a recruiting
film and their Quarterly Magazine –
this issue:"Helping the Arts escape red ink."
Illustrator, Reynold Ruffins. 1968

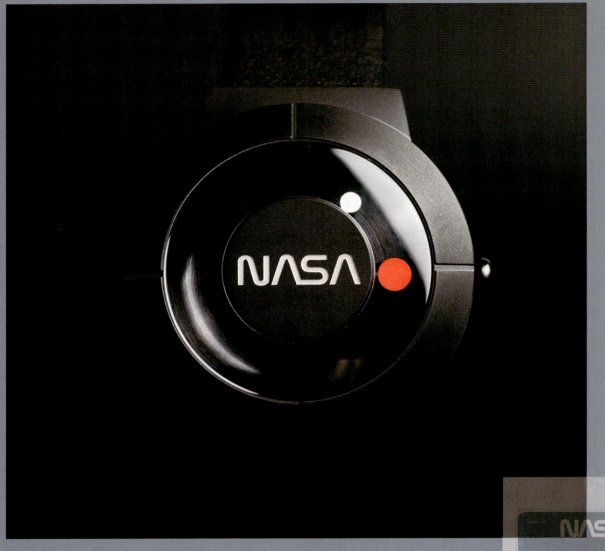

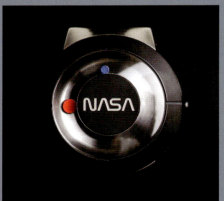

The Space Watch:
Homage to a 50-year relationship with
NASA, and designed to reflect space
itself – its silence, its revolving planets.
Watch commissioned and produced by
Anicorn Watches. 2021

I've been privileged to lead our design institutions and help elevate the profession over transformational years, and I'm delighted to pass on this career knowledge.

Collections & Podcasts

Collections, Interviews & Books

Richard Danne has lectured to a great variety of audiences and at diverse locations both nationally and internationally. He also taught for several years at SVA in New York City. Items below are available online:

AIGA "Inspiration Collection"
Featuring some 150 design projects. On the AIGA site, go to "Design Archives" then "Collections," and scroll down to "Richard Danne," click on to view. With special thanks to Ric Grefé, former executive director of AIGA.

Graphis "Masters" Collection
A collection that encompasses six decades of design in 30 projects. Thanks to B. Martin Pedersen of *Graphis*, the premier publisher of design.

AIGA/SF Fellow Award
An expansive interview by the noted Karin Hibma, on my design career and professional leadership, for the 2018 AIGA San Francisco Fellow Award.

Designculture
A personal, penetrating interview with Nicola Munari about my life in design over the decades (with a career-tracking portfolio).

Rob Auchincloss: HOLOCENE
An extensive, far-ranging interview hosted by this prime space / design thinker.

Standards Manual
The NASA Graphic Standards Reissue, from the impressive enterprise of partners Jesse Reed and Hamish Smyth of the design firm *Order*:

"Illuminations"
Collection of personal photographic images from around the world exploring the effects of "Light + Shadow."

Podcasts & Videos

Most collections and
podcasts are available online
by simply Googling:
Richard Danne, designer

Richard has also participated in numerous podcasts, some focusing on the
reintroduction of the NASA Logotype into their program and vocabulary.
Here are a few of those casts:

Town Hall

For all NASA employees, hosting and interview by Bettina Inclan, deputy
director for Communications. With Bert Ulrich of NASA Headquarters and
David Rager of JPL doing the new Design Guidelines presentation.

"Houston We Have a Podcast"

A radio podcast from Johnson Space Center, hosted by Gary Jordan for
all NASA employees. "After 28 years, the NASA "Worm" is back and had a
high profile reintroduction during NASA's SpaceX Demo-2 launch in May 2020."

Strand Bookstore Conversation with Michael Bierut

A spirited, informative dialog tracing the ups and downs and ups of the iconic
NASA Graphics Standards Program.

NASA Graphic Standards Manual

A video created for the record breaking Kickstarter campaign from *Order*.

San Francisco Clarity Conference

Keynote address for this inaugural conference of UX designers.

Brian Collins, COLLINS Agency

For all Collins employees and some 150 other invited designers from around
the world. A young audience.

Shopify "Summer Sessions"

A major presentation for all 5,000 UX employees of this thriving enterprise.
(not for external viewing)

Richard Danne AIGA Medalist

An introductory video at the awards ceremony for the Centennial Medal.

"Ask Me Anything"

An intimate cast with Jina Anne, the creator of the Clarity Conferences.

AIGA / SF Medalist Presentation

A comprehensive video produced and hosted by Karin Hibma for the AIGA/
SF chapter during National's Centennial Celebration.

ANXIETY

Richard Danne
b. 1934

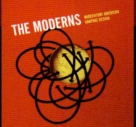

THE MODERNS
MIDCENTURY AMERICAN
GRAPHIC DESIGN

STEVEN HELLER
GREG D'ONOFRIO

NASA

"The Moderns"

When you're young and trying to envision what your life might look like, you can be forgiven for dreaming. As a youngster on that Oklahoma farm, I needed a vivid imagination to lift me from the sod, so as to fly.

In late 2017, the noted design historians, Steven Heller and Greg D'Onofrio, introduced their landmark 336-page book entitled "The Moderns," published by Abrams. It features "designers working from 1937 – 1970, who invented and shaped Midcentury Modern graphic design in America." Some migrated from Europe, while others were born here and were "home grown."

So, this is far beyond my boyhood dreams. I've been blessed to achieve much in a wonderful six-decade plus career – it's been an exhilarating ride! I'm honored to be included in "The Moderns," this rich resource of *Timeless Design* for future generations.

"Meet The Moderns" book launch event, November, 2017, NYC.
Greg D'Onofrio, moderator (left) with Moderns:
George Lois, Rudi Wolff, Tom Geismar, Richard Danne

1936
Our hired thrashing crew, with
horse-drawn wagons.

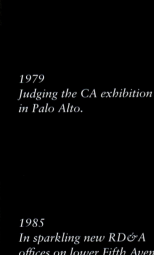

1944
After surviving a critical
month-long stay in the hospital.

1955
Performing in Colorado Springs
with our modern jazz quartet.

1979
Judging the CA exhibition
in Palo Alto.

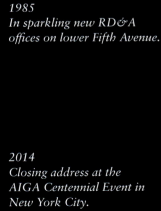

1985
In sparkling new RD&A
offices on lower Fifth Avenue.

2014
Closing address at the
AIGA Centennial Event in
New York City.

About the Author

I was born in 1934, in the middle of The Great Depression, and Dust Bowl Oklahoma wasn't the most welcoming of places. Times were tough on the farm – drought conditions in the mid-western states had created a debacle of immense proportions and all was compromised by the punishing weather in the mid-thirties in middle-America. It lasted for a very long time!
Richard Danne (DAY-nee) from his memoir "Dust Bowl To Gotham."

I believe that Richard's hardscrabble early years during the Depression and Dust Bowl made him who he is today. These were very difficult times and people needed to help each other, and he learned from that and became dedicated to service. Without his national leadership, voice, and vision, AIGA chapters like ours wouldn't be here today.
David Hisaya Asari, AIGA/SF President, Centennial Medalist event introduction, 2014.

As much as any other individual, Richard was responsible for transforming AIGA into a national, chapter-based organization.
Michael Bierut, partner Pentagram/NY at the Strand Bookstore interview, New York.

In my 30 years in education, I have never encountered such an intelligent and intuitive designer who consistently exceeds expectations.
Maureen Brenner, then Head of Riverview School, East Sandwich, Massachusetts.

Inspiring. Your commitment and integrity are timeless, alway refreshing and important.
Jack Anderson, Chairman, SidLee, commenting on a recent Italian interview, Seattle.

So cool! Is there any limit to this guy's talent?
Robert Pizzo, noted illustrator comments on Richard's 2022 piano jazz album, New York.

Best possible ending to Clarity 2016: hearing from Richard Danne on the NASA Graphics Standards Manual.
Trish Ang, UI Engineer tweeting at the first Clarity Conference, San Francisco.

Thanks for your dedication in helping make this wonderful museum a reality.
John Hightower, then executive director, South Street Seaport Museum, New York.

Congratulations, Richard. You've had a fantastic career! And, you being one of the fathers of the modern AIGA and a Centennial Medalist, I am delighted that you have been named an AIGA/San Francisco Fellow.
Karin Hibma, : : CRONAN : : partner, during an interview for the 2018 Fellow Award.

Former president of AIGA, Dick Danne is in the battle of his life. We are all rooting for him to get through a tough period, a health crisis. So, on the count of three, I want you all to shout "We Love You Dick Danne!" And we'll send the video on to him.
Dana Arnett, AIGA president, at the 2018 National Conference, Pasadena.

So, I'm wondering, why aren't you more famous?
Rob Saunders, Executive Director & Curator, Letterform Archive, while selecting projects for a Danne / Archive collection, San Francisco.

Richard, oh my!! Thank you so very much on behalf of Brian, the COLLINS team, and everyone you've inspired today. The feedback is just out of control: people from all over the world joined the talk; we had guests from Mexico, England, Russia, Germany, South America, and everyone left with a big smile and a big gratitude and inspiration!
Katya Braxton, director, Resourcing @ COLLINS, after their 2020 Podcast.

Special Thanks

Now comes the pleasant task of expressing gratitude to those fine people who have contributed so much to my rewarding career (long career = long lists).

Clients and Sponsors

Gloria Kane and James Bowers – for their early and enthusiastic support in Dallas.
Sheldon Cotler of Time International – for helping launch my free-lance practice in NYC.
Jonathan Rinehart – we did a lot of fine work together at Adams & Rinehart, and Ogilvy.
Nancy Yedlin and Loretta Keane of FIT – Yes, Extraordinary!
Also: Charley Rockwood, Edgar Bronfman, Sr., Bob Kristan, Alec McGhie, Faith Wohl, John Steeves, Jim Cunningham, Langhorne Bond, Betsy Halpern, Dr. Mack Lipkin, Jr., Marilyn Laurie, Dr. James Fletcher, Jack Bailey, Chris Conover, Jim Foster, Dennis Dinan, John Hightower, Bill Powers, Jerry Perlmutter, Jeanne Krause, J. Murillo Valle Mendes, Melinda Hamilton, John Solomon, Dr. William Banko, Keith Anderson, Stan Scott, Toby Lineaweaver, Barbara Field, Tad Ware, Murray Grant, Chris Danne, Jerry Blood, Alex Wellins, Dick Gray, Walter Allner, Ron Campbell, Irwin Grossman, Steve Frankfurt, Mark Roberts, Vito Abrietis, Harry Levine, Peter Abbott, Jake Jacobi, Bob O'Brien, David Brown, Howard Chapnick, Kathleen Ritch, Lee Geist, Chuck O'Rear, Lisa Sheehy, Ed Russell, Maury Bates, Brian Lacey, Sid Rapoport, Don Davidson, Keith Kramer, Robert Taylor, Peter Borelli, Hugh Kaiser, Joe Kwan, Maureen Brenner, Rich Delaney, Milton Ferrell, Rob Saunders, David Fitch, Burt Ulrich, and my publisher Gordon Goff.

Partners + Staff

Deep thanks to partners: Philip Gips of Gips & Danne / 1964-69, and
Bruce Blackburn of Danne & Blackburn / 1973-85.
Also to stellar staff: David Griffing, Stephen Loges, Gary Skeggs, Don Meeker, Carlos Serrano, Bob Silverman, Marlowe Goodson, Juliet Shen, Kevork Babian, Bruce Johnson, Diana DeLucia, Reid Martin, Kurt Jennings, Philip Goldberg, Margaret Wollenhaupt, Vera Chazen, Charles Unger, Maria Kerdel, Dr. Melinda Knox, Renae Baer, Gayle Shimoun, Andrea Barash, Klieber Santos, Shaun McDavid.

Design Colleagues

Through their work or friendship or both: George Tscherny, Paul Rand, Lou Dorfsman, Milton Glaser, Seymour Chwast, Saul Bass, Lou Danziger, Rudy DeHarak, Erik Nitsche, Gene Federico, Henry Wolf, Jim Cross, Pierre Mendell, Tom Geismar, Ivan Chermayeff, Tom Gonda, Ken Parkhurst, Ben Bos, Lella & Massimo Vignelli, Brad Thompson, Fritz Gottschalk, John Massey, Steff Geissbuhler, Herb Lubalin, Henry Steiner, Bruno Uldani, Ruedi Ruegg, Linda & Kit Hinrichs, Leo Lionni, Alan Fletcher, B. Martin Pedersen, Colin Forbes, David Goodman, Stuart Ash, Bart Crosby, Takenobu Igarashi, Paul Davis, Bruno Monguzzi, Tomoko Miho, Jim McMullan, Sheldon Cotler, Dick Hess, Sean Adams, Michael Vanderbyl, Ken Carbone, Leslie Smolan, Steve Heller, Stan Richards, David Hillman, RO Bleckman, Jennifer Morla, Dana Arnett, Ken Cato, Jack Summerford, Chris Pullman, Jackie Casey, Niklaus Troxler, Stephan Bundi, Michael Bierut, Louise Fili, Jim Miho, Greg D'Onofrio, Woody Pirtle, Paula Scher, Harry Diamond, Michael Donovan & Nancye Green, Sam Antiput, Tony Russell, Ruth Ansel, Andy Kner, April Greiman, Alexander Isley, Ray & Charles Eames, Eric Baker, Arnold Saks, Brian Collins, Jesse Reed, Hamish Smyth, Calvin Woo, Susan Merritt, Jim Stockton, Clement Mok, Lyle Metzdorf, Elton Robinson, Jack Anderson, Roger Cook, John Follis, David Asari, Arnold Shaw, Bob Appleton, Mary & John Condon, Bob Scudellari, Lars Muller, Brad Holland, Armando Milani, Robert Pizzo, Dan Goods, DK Holland, Keith Goddard, Roger Whitehouse, Richard Poulin, Kelly Brandon, Lana Rigsby, Michael Cronin, Jennifer Sterling, Glen Cummings, Martin Donald, Phil Hamlett, Pat Taylor, Craig Frazier, Chris Hill, Bob Vogele, Jim Sebastian, Lance Wyman, John Bielenberg, Cathy Hull, Harold Burch, Art Shipman, Earl Gee, Anthony Russo, Jessica Helfand, John Van Dyke, Kelly Brandon, Joel Katz, Mathew Carter, Derwyn Goodall, Jon Craine, Jane Brown, Bill Bonnell, Barbara Balch, Bill Caldwell, Jennifer Kinon, Peter Good, Ken Bess, Katherine McCoy, David Renning, Robert Burns, Willy Kunz, Laura Scott, John Rieben, Rex Petite, Don Sibly, Michael Mabry, Anthony Simms, Hugh Dubberly, David Rager.

Craft / Media

Dr. Robert Leslie, Dick Coyne, Martin Fox, Peter Lawrence, Ed Russell, Milton Kaye, Gordon Kaye, Ed Katz, Stan Scott, Bob Ott, Sr., Fran Canzano, Doug Bittenbender, Sid Rapoport, Aaron Burns, Norman Sanders, Don Davidson, Howard Levy, Allen Johnston, Joe Messina, Bill Berger, Sarah DeWalt, Fred Salazar, Ken Friedland, Tad Crawford, Stu Leventhal, Andrew Wilson, Ben Chapnick, Patrick Coyne, Jean Coyne, Kenneth Chang, Christine Reid.

Beyond Category

Caroline Hightower, Ric Grefé, Julie Anixter, Ed Gottschall, Shannon Feeney, Chris Danne, Karin Hibma, Deborah Brown, Mike Schacht, Nancy Hanks, Chris Ricchi, Joe Messina, Marcelline Thompson, Stowe Phelps, Dorothea Hutton, Ralph Caplan, Suzanne Slesin, Silas Rhodes, Betsy Halpern, Jina Anne, Vicki Greene, Rose DeNeve, Kathy Shorr, Margarethe Laurenzi, Bert Jackson, Bob Schulman, Paul Duffy, Susan Marks, Allan Stanley, Hal Lindgren, Ferg Harvey, Ron Williams, Sam Olkinetsky, Elinor Evans, Dale McKinney, Bill Farrell, Sue Cross, Dick Missan, Florence Wettig, Nathan Gluck, Harris Danziger, Bert Benkendorf, Debbie Millman, Dawn Zidonis, Stacey Panousopoulos, Gabriela Mirensky, Rob Auchincloss, Nicola Munari, Mike McNulty, Dennis Moss, Heather Strelecki, Bert Ulrich, Elena Sansalone, Amy Chapman, Kaelig Deloumeau-Prigent, Jim Dean, Chris Larkin, Linda Peek Schacht, Dr. Gjoko Muratovski, Robin Landa, Larry Kevy, Gloria Kondrup, Hillary Coe, David Doody, Bettina Inclan, Alex Daly, John Solomon, Danne Woo, Kirby Anderson, Dr. Babis Andreadis.

Credits

Book Design
Richard Danne

Project Design
Richard Danne was creative director for the projects
in this book, and designed most individual pieces.

Several talented staff designers made contributions to
specific projects and they are:
Eric Atherton: pages 56 (marquee), 156 (bag),
171 (Atlantic), 205,* 212,*213.*
Renae Baer: page 149 (annual).*
Andrea Barash: page 148.*
Bruce Blackburn: pages 16-19 (original presentation),*
20 (logotype).*
Martin Donald: pages 77,* 138.*
Phil Gips: pages 138 (mark), 154-155* and 200.*
David Griffing: pages 115 (mark), 170 (love), 199.*
Kurt Jennings: page 60.
Maria Kerdel: page 52 (apparel), 55.
Stephen Loges: pages 22,* 54 (catalogues).
Donald Meeker: pages 145-147.*
Gale Shimoun: page 57 (toys).
Gary Skeggs: pages 22*, 48, 52 (marks), 53 (media kit),
54 (*Network* format), 128, 130-131*, 133 (manual),
149 (mark), 150, 152-153, 157 (exhibit), 158,
161 (manual), 185 (business),* 216 (lobby banners).
Charles Unger: page 167* (YPO).
Margaret Wollenhaupt: page 205.*

Designed with Richard Danne

Typography
Text set in Sabon LT Pro.
Headings and folios set in Meta Pro in various weights.
Cover set in Franklin Gothic Bold Condensed and
Sabon LT Pro Italic.
Typesetting by DanneDesign, Napa, California.

Design & Production
Created in *QuarkXpress*, then imported to *InDesign*
for proofing, production, and offset printing.

Art
Many of the images in this book are from the pre-digital
era. They were scanned from 35mm or 4x5 film,
or from actual printed samples (please be kind). Images
from post-1994 are from the original digital files.

Photography

Numerous photographs courtesy of NASA:
Covers: *Uranus*.
End leaves and single pages: *Abell* galaxies in the constellation *Draco*.
Forward page: *Andromeda* constellation.
Page 12: Space walk.
Page 19: Earth rise by Astronaut William Anders.
Pages 20-21: All photos.
Page 26: Logo on the *SpaceX* rocket by former NASA Administrator Jim Bridenstine.
Page 27: *Orion* spacecraft and Logo sculpture.
Page 30: Headquarters Events.
Page 80: Earth image.
Page 225: AIGA Gala.

Featured projects:
Herman Bachmann: pages 75, 207, 231 (*Review*).
Jim Barber: pages 49, 52, 180 (*Connections*), 183, 186, and design portfolio studio still-life photos.
Rene Burri: pages 19 (launch photographers), 105-110.
Chris Callis: page 212.
Richard Danne: pages 90 (cover), 93, 102, 121, 234.
Center For Coastal Studies: pages 88-89, 94.
Gabriel Amedeus Cooney: page 85 (*Voice*s and annual).
Wayne Eastep: pages 192, 194.
George Elliott: page 155.
Bill Farrell: pages 116-117, 149, 228 (*ABS*).
Paul Fusco: pages 210-211.
Taishi Hirokawa: page 60.
iStock: page 124, McCaw image.
Magnum photographers: 2 photos per member, page 177.
Phil Marco: page 226 (*Tafex*).
Nancy Moran: pages 62-63 (clouds), 66-69.
Alan S. Orling: pages 14, 118-120, 145-147, 156, 208, 211, 216, 219, 230.
Jeff Perkell: pages 176, 217 (both *Seaport*).
John Senzer: pages 47, 50-51, 59 (*LookBook* photos).
Cynthia Stern: pages 22, 206, 213.

Other photography

Partial list of photographers for other projects includes:
Burk Uzzle, Costa Manos, Burt Glinn, Alex Webb, Elliott Erwitt, Jay Maisel, Bruce Davidson, Getty Images, John Madere, Steve Langerman, Bob Colton, iStock, Chris Maynard, Al Giess, Gloria Baker, Kelly-Mooney.

Illustration

Credit appears with its image.

NASA Odyssey story title inspired by the 1968 Stanley Kubrick film *2001: A Space Odyssey*

DanneDesign
16 Cove Court
Napa, CA 94559

e mail@dannedesign.com
w www.dannedesign.com

Published by ORO Editions